Solidworks Simulation for Real Machines

Copyright 2020,

Stone Lake Analytics, LLC

Seneca, South Carolina

www.stonelakeanalytics.com

ISBN 978-1-7335955-5-1 (paperback)

ISBN 978-1-7335955-6-8 (e-book)

Contents

Introduction

Target Audience

This book is aimed at people who are ready to build a machine, starting now, or repair a broken machine, starting yesterday. By "machine" we mean a thing with multiple essentially rigid parts which go together to do a mechanical job. The parts may roll, slide, or spin on each other, or they may be bolted or welded together.

The machines with which we have the most experience are in the business end of material handling equipment, like fork lift masts or telehandler attachments. Over some years we have developed techniques to handle the parts of these products, like chains-over-sheaves or paired hydraulic cylinders, in basic simulation tools. Here we show how each element is handled for both accurate results and quick setup and run times.

Engineering managers, the kind who used to spend hours flipping through a McMaster-Carr catalog looking for parts to keep a project going, will learn from this book just what can be done with only fundamental FEA simulation tools. Anyone buying analysis services will similarly get an idea what to expect, or demand, from a competent outside service. This book is a demonstration of Stone Lake Analytics' capabilities.

Users of Solidworks Simulation will not get click-by-click instructions on how to set up the studies shown here. But users of any contemporary FEA tool should get both an idea of what is possible using such a tool, and a head start on setting up small scale studies which can lead to full-fledged high-accuracy models of whatever they are trying to make.

Why Solidworks Simulation?

Solidworks is what Stone Lake Analytics chose to work with, because of the high integration of tools at an attractive price. That has always been the core motif of Solidworks sales. For simulation we think the solid modeling features and configuration management make Solidworks the very best pre-processor available.

Anyone who has kept up with the CAD/CAE software market is probably aware of the software tools that Dassault-Solidworks has acquired over the years. The user interfaces of Solidworks Simulation are pretty sparse, sometimes frustratingly so, but there is some seriously powerful code underneath. The patient user can tease out the best parts of the code by patient testing, trying different treatments of different parts, until he or she is ready to put everything together and run a complete machine.

Hardware Considered

Following is an incomplete list of the machine elements we will consider and illustrate.

Plain bearing
A pin passed through a number of round holes makes a common plain bearing. We can represent this very simply with Solidworks Simulation connectors or very literally with contacting solids.

Bushed Joint
Around some plain bearings designers will put different types of bushings. We can add them to the bearing and get explicit data about forces on the bushings.

Ball Joint
Ball joints (pictured) can also be represented number of ways, allowing intended free motion while also getting force data at the joint.

Slider
Simple sliding wear pads are often used to take up secondary loads. We can include them and get design data like P-V values.

Roller
Rollers also can be modeled literally as moving parts, or as sliding solids.

Fluid (hydraulic/pneumatic) Cylinder
Cylinders (pictured) bring fluid forces into a machine from outside the assembly, allowing free motion in some directions while locking total length.

Cylinder Pair
Paired cylinders set up special kinematic relationships using fluid outside the mechanical chain. We will demonstrate several ways to account for this while keeping forces balanced correctly.

Chain/rope Over Sheave
Chains and rope can be loaded only in tension. Sheaves pair up different lengths at the same force, which we will capture.

Die Sets
Tight fitting stacks of solid blocks come together to make stamping and casting dies. We can put them in contact and look at die performance in fine detail.

Bolted Flange
Connecting parts with bolts sets up a sometimes surprising load pattern in the mating surfaces. We can get the complicated stress fields here, with some necessary expense.

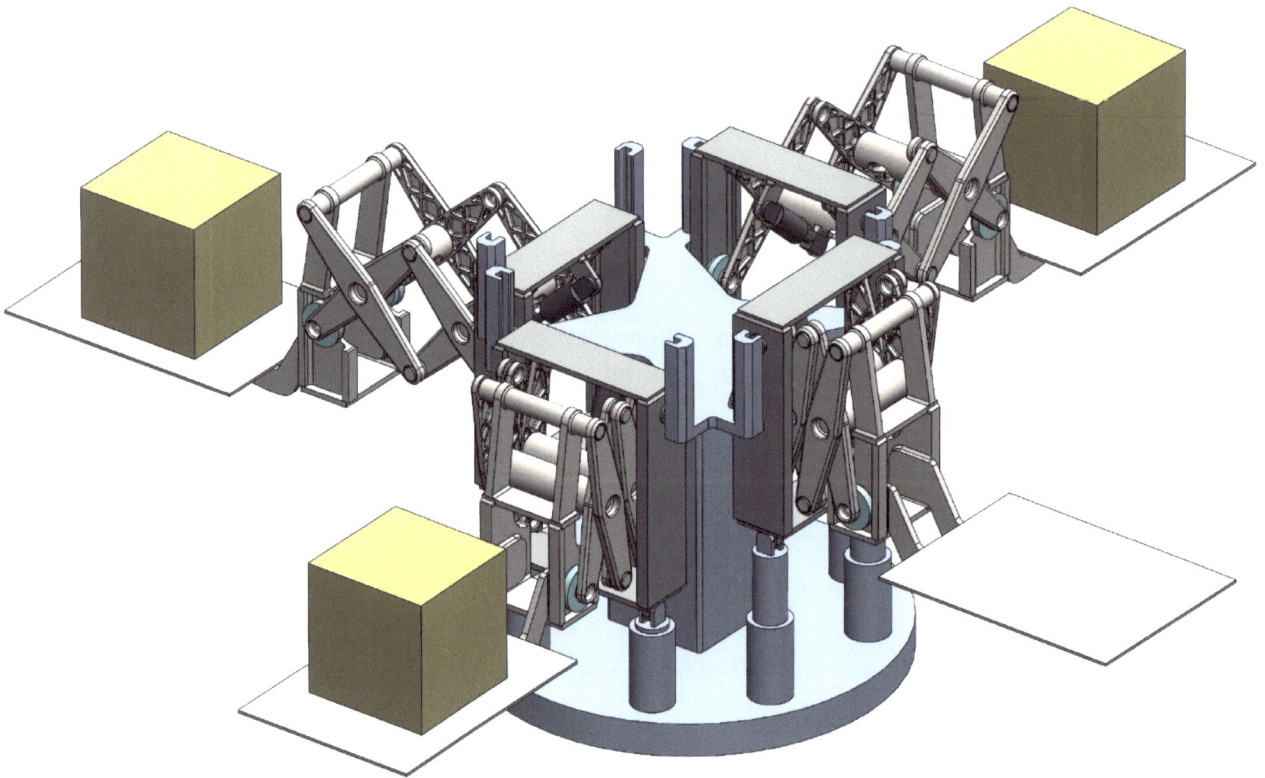

Carousel with four double-pantograph platform lifts – includes roller bearings, bushings, linear sliders, and hydraulic cylinders

Scope

Any good analysis is rooted in a well-posed question. The objective is to answer the question, usually with a yes/no (pass/fail) or quantitative result. The objectives of the studies we will build up to here are:

- Stress prediction in structural elements
- Force prediction at hardware which can be compared to catalog ratings

The internal objective of the studies is to get these answers in a reasonable amount of time. What is reasonable depends on the complexity of the machine under study; what is useful depends on the current schedule. We think the vast majority of jobs can be set up by an experienced user in one morning and results be in hand that afternoon, using a common desktop workstation.

Linear Static Analysis (Snapshots)

Everything done here is with linear static analysis, the most basic FEA solution. There are much more complicated simulation tools available, which can capture dynamic events and even catastrophic collisions, but they require much greater computer and analyst effort. We think most machines must first be designed for 'normal' usage, and that smashing into a wall is usually outside the design brief[1]. In Solidworks Simulation getting a good static solve is usually the first step to set up of other study types.

When we set up an analysis for a linear static run, baked into the process are certain assumptions:
- Nothing broke
- Nothing yielded
- Nothing moved (very much).

Since avoiding component failure or yield of ductile material is the usual goal, we must check the validity of the solution, but this is not hard.

Even a machine with a great range of motion is, and should be, designed with certain "snapshot" conditions framing the design. So, we pick positions of greatest interest to set up static studies. These will typically be the expected worst-case loadings, e.g. fullest extension, greatest load, or highest acceleration.

Quasi-Dynamic (inertial) Loading

Acceleration? Yes, acceleration can be included in a static analysis. A constant acceleration, like gravity, can be applied to any machine. We will usually include gravity and here we will look at cases with linear acceleration (a truck-mounted machine when the truck is stopping) and a rotating reference frame (a machine mounted to a large turntable).

Conditions for Using Linear Static Analysis

Sticking with linear static analysis means that, for well-posed problems, solutions with a good solver and modern computer can be lightning fast. We think it's a good choice for most design problems, but certain assumptions must be kept in mind.

By "linear" we mean that materials are expected to be elastic, with a linear stress-strain relationship. This is a very good assumption for common metals, as often used in durable designs. ONLY the elastic (Young's) modulus and the Poisson's ratio of the material are inputs to a linear solution. (This will have to be explained to clients and co-workers *ad nauseum*.) In the studies here generic alloy steel and wrought aluminum are used interchangeably with named alloys. Other material properties, e.g. yield strength and hardness, come into play when evaluating results but not in the setup.

The other catch to a linear solve is that deflections cannot be too severe. Everything should start out close to where it's going to wind up. This keeps the setup inputs correct for the final shape and can

---

[1] This has not precluded our lead analyst from working as a live crash test dummy to prove out a machine.

greatly speed up contact solutions. This is not a shortcoming if a design objective was limited deflection – big bending can be a failure by itself, irrespective of stress and force predictions.

By "static" we mean that the machine is not moving or is moving at constant acceleration. This covers a lot of applications perfectly well and is a baseline for more complicated studies in any case. Also remember that the very definitions of properties like yield strength depend on slow application of forces. Truly dynamic behavior requires a different way of looking at materials, not just a long simulation solve time.

Contact

Contact between parts is key to much of what we do. Solidworks Simulation offers a number of ways to handle parts that may or may not touch, from easy (put everything into no-penetration contact) with a long solve time and difficult convergence, to tedious (manually assign every contact set) but with a faster solve and lots of useful data. We will look mostly at how to do the latter, with minimum user time and most useful output.

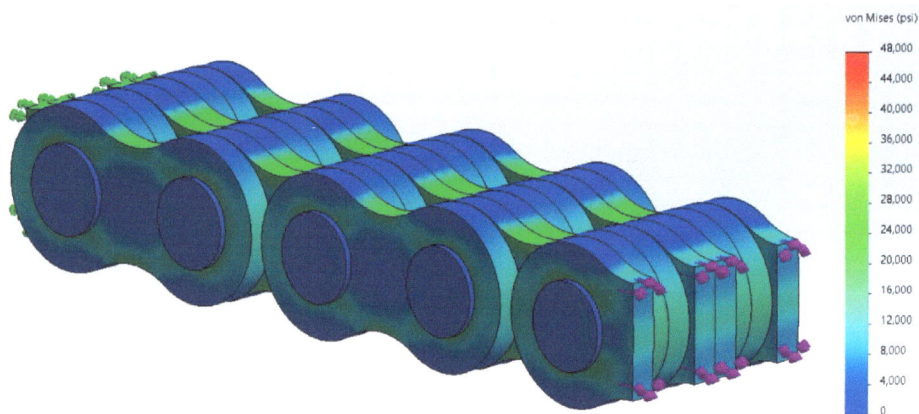

Multi-part leaf chain simulated under tension in full contact

Connectors

Component interactions often happen through pieces of purchased hardware, like bearings or chains. Solidworks Simulation offers a number of special "connectors" which can fill in for these catalog components. After solution force data is provided for the connectors which, if the setup is right, can be compared directly to manufacturer ratings. Connectors can also be used to set up kinematic relationships between parts, avoiding the need to model parts which may be a long way from final design. The "bolt" connector is a special and powerful tool.

Impact of Connectors

The obvious benefit of using connectors is that a complicated part, like a roller bearing with every pin, cage, race, and seal modeled in fine detail, can be replaced in the model by a single entity [apologies to the bearing makers who put such great models on their web sites]. Repetitive assemblies like a leaf chain can similarly be swapped out. The first consequence of this replacement is a loss of fidelity; some thought must be put into what connectivity and flexibility is being represented. The next consequence is that the model no longer looks like the real deal. This is an issue when it comes to presenting results to others – a little more computer time may be worth not having to explain things repeatedly.

Point Loads

The easiest replacement for a chain or rope is a "link" connector. A force relationship can be tracked well this way, but the connection on either end must be simplified and there will be point loads on either end of the link. The resulting stress hot spots will have to be explained and stress on the real connection geometry evaluated separately.

A "spring" connector can be used for the rope or chain, even a tension-only spring should there be a possibility of the chain going slack, but now we are adding complication and solver time back in. We will see an example later.

Local Stiffening

Internally most Solidworks Simulation connectors generate a lattice of hidden elements to achieve the desired effect. Sometimes these elements are perfectly rigid. So, the connected surfaces might be artificially constrained on a surface, like the internal cylindrical surface where a bearing is to be pressed. Also, forces on that cylinder might be spread all the way around, when in reality the bearing can only press on the race in one direction at a time.

We should note right here that in recent Solidworks versions behavior of the "pin" connector has been greatly improved in this regard. We will explore this in detail.

Boundary Parts

Much mental effort and worry is spent over proper setup of boundary conditions. We have a simple answer – don't! We recommend putting all boundary conditions a component away from the actual parts of interest. Once simulation with contact and connectors is understood, one can quickly model interfacing mounting or equipment. Then the part or assembly being studied can be bolted or bonded or pushed up against something with realistic compliance. This need not take a lot of time.

Interference

Solidworks Simulation allows parts to start out with some interference, then it can calculate forces needed to overcome the interference. The interference can be deliberate, like a press fit, or a hypothetical problem, like captured debris. An example of trapped die flash appears later.

Studies

Study 1: Bolted Post, Bolts in a Square Pattern

Overview

In this case study we will work with a single solid part, a square tube post welded to a flat square base (The parts are joined, welded, and gapped as described in Solidworks Simulation for Real Weldments). We will examine boundary condition options, bolt connectors, and use of mating bodies instead of direct boundary conditions.

Objective

The key results from this study are prediction of peak stresses in the weldment and forces in the bolts.

Challenges

We must find a setup which accurately represents the mounting without undue use of human or computer resources.

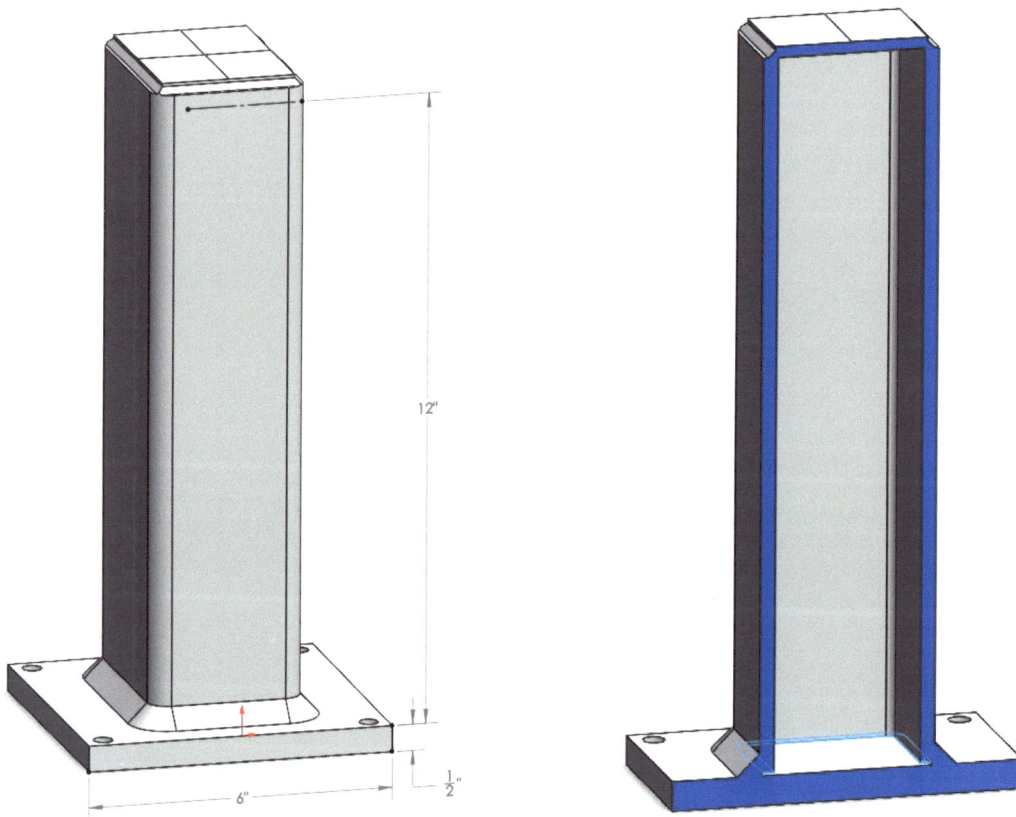

Solid model, dimensions and section view

Square Pattern, Fixed Base

The post is loaded with side shearing forces at the top. For a first pass analysis the base is fixed at the bottom face.

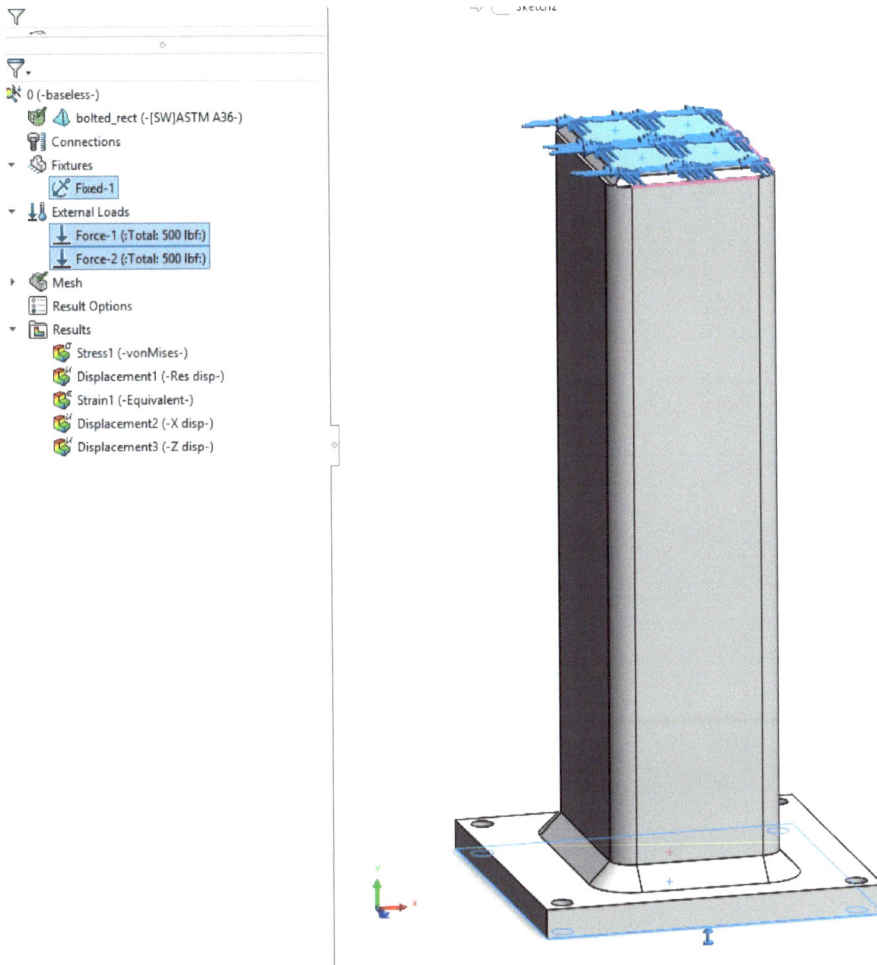

First simulation setup

In simulation, this geometry is identical to the result if one ran the assembly with bonded contact, treating the mating surface as a union. So we can set up the first simulation with this shape directly.

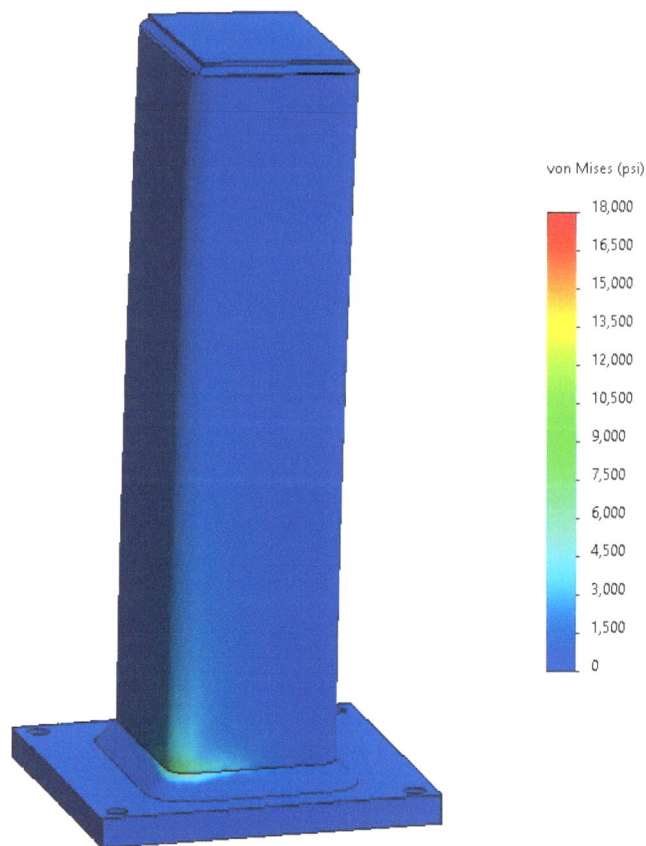

Simulation result (150X displacement)

Response in the post and the welds looks 'reasonable'. There is bending in expected directions and peak stress is at the top of the weld bead. But there is no apparent stress in the mounting plate. Since the plate is completely fixed on one side, this is no surprise.

We want to know what is really happening in the plate, and if it affects the weld. So the setup is improved, at first without any additional solid modeling. A "virtual wall" fixture is set up. This is a rigid plane, or imaginary hard solid body. The modeled part will be in contact with the plane; so our part can pull away from the plane but cannot pass through it.

Square Pattern, Virtual Wall

The next step is to define "foundation bolts" through the four holes. These special connectors require the virtual wall and the contact condition. They are drawn by Solidworks as a version of J-bolts that are sometimes cast into concrete for just this purpose. Specific anchor dimensions and nut pre-load (torque or force) can be input.

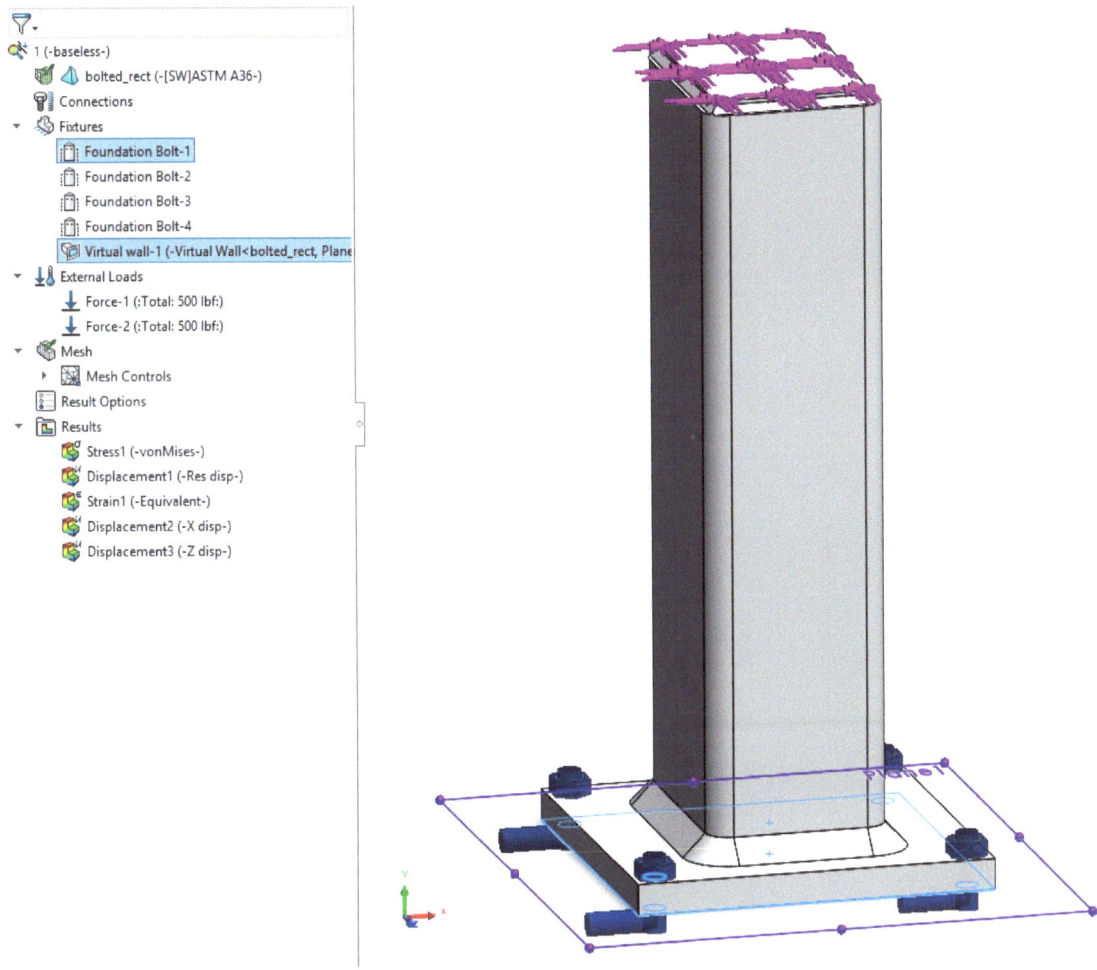

Setup with virtual wall and foundation bolts

The stress in the post is not changed but total displacement is greater and stresses in the plate and weld are dramatically different. Sharp stress peaks are seen at the weld edges.

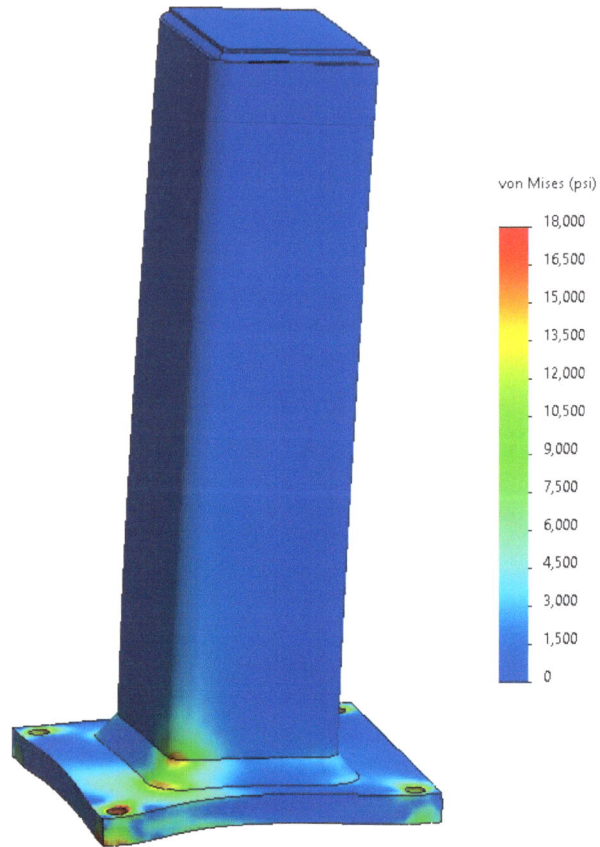

Result with foundation bolts

The base plate is now allowed to flex. It pulls away from the reference plane, restrained only at the bolt holes. The maximum stresses are seen at local constraints, fasteners and the weld. This is how it goes for most machines we encounter. A traditional simple boundary condition would never reveal this.

Square Pattern, Solid Base

Setting up the virtual wall and bolts was a few extra steps though. It may be just as easy, and more clear to others, if the post was bolted to a piece of visible solid material.

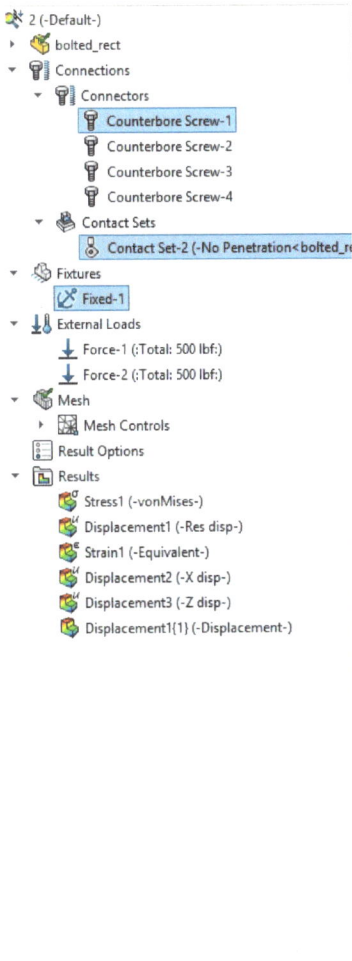

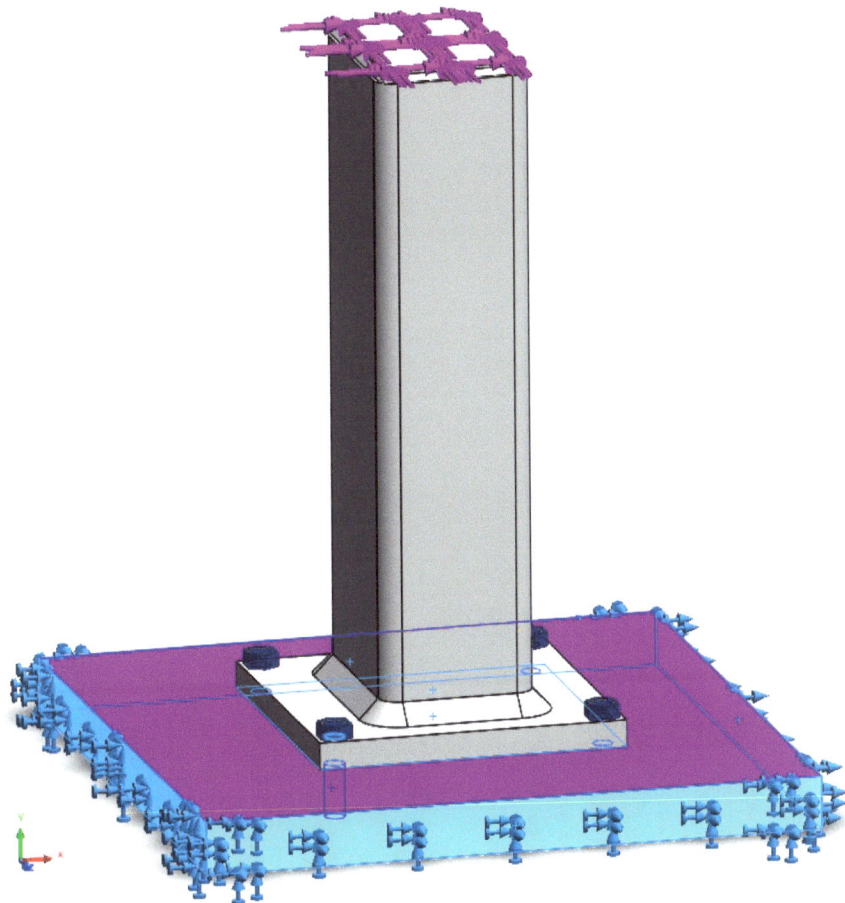

Setup with solid base and bolts

A piece of material was modeled below the post in one step, the bolt holes being copied from the post. Bolt connector creation now consists of picking two cylinders for each instead of a cylinder on the part and the rigid reference plane. The sides of the new base are fixed all around. The base can be made of the real base material (e.g. concrete or steel plate) or just anything relatively stiff compared to the part. In this case it is the same material as the post (mild steel).

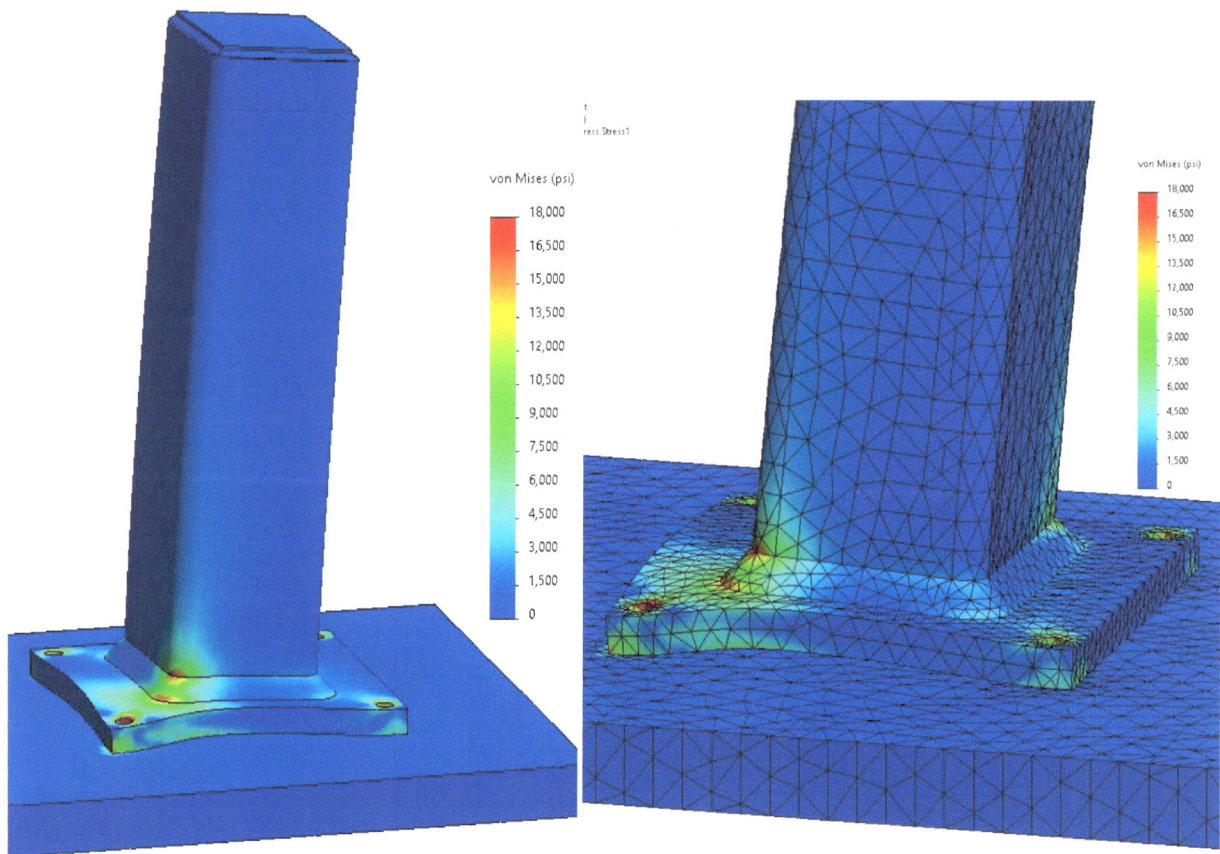

Results with solid base and bolt connectors

Stress results are practically identical to the virtual wall study. The solid base has some compliance, but it is much thicker than the mounting plate. A plot with mesh edges shows the differing mesh density between the plate and base, which can be fairly coarse. Adding the base does add to the total mesh size and calculation time, but we expect if you try it both ways you will have a hard time measuring the difference [of course running with the simple fixed boundary condition was extremely fast – it also gave bad results!]. We think adding the boundary solid is harder to mess up and easier for other people to understand.

We are not done with the study when we get stress and deflection. Because we defined discrete connectors, we can ask for data from each of them. Here we get forces on the hardest working bolt.

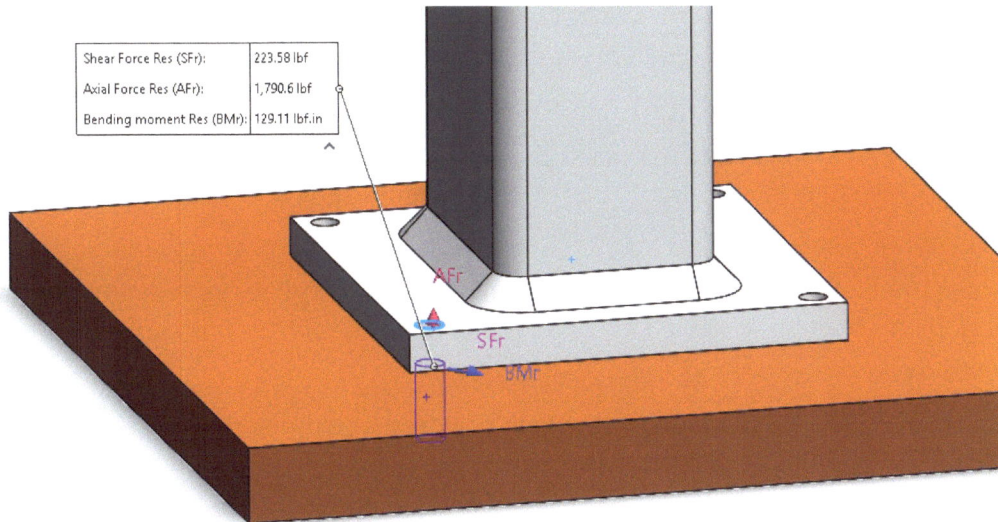

Shear Force Res (SFr):	223.58 lbf
Axial Force Res (AFr):	1,790.6 lbf
Bending moment Res (BMr):	129.11 lbf.in

Force data from bolt

We have looked at a simple application with a rectangular bolt pattern. This is usually a sound attachment method. Now we see how fasteners can be used badly, all in a single line. We will also look in more detail at what happens in a bolt connector.

Study 2: Bolted Bracket, Bolts in a Linear Pattern

Overview

This part is another steel weldment. It is meant to carry some load, a heavy spool perhaps, and be bolted to a vertical wall. The same mounting types as in the previous example are used

Objective

The key results from this study are prediction of peak stresses and relative deflection in the weldment. . The option of modeling bolts as literal solids is to be explored.

Challenges

We expect the majority of the part's mating surface to pull away from the wall. This will invalidate common boundary condition assumptions.

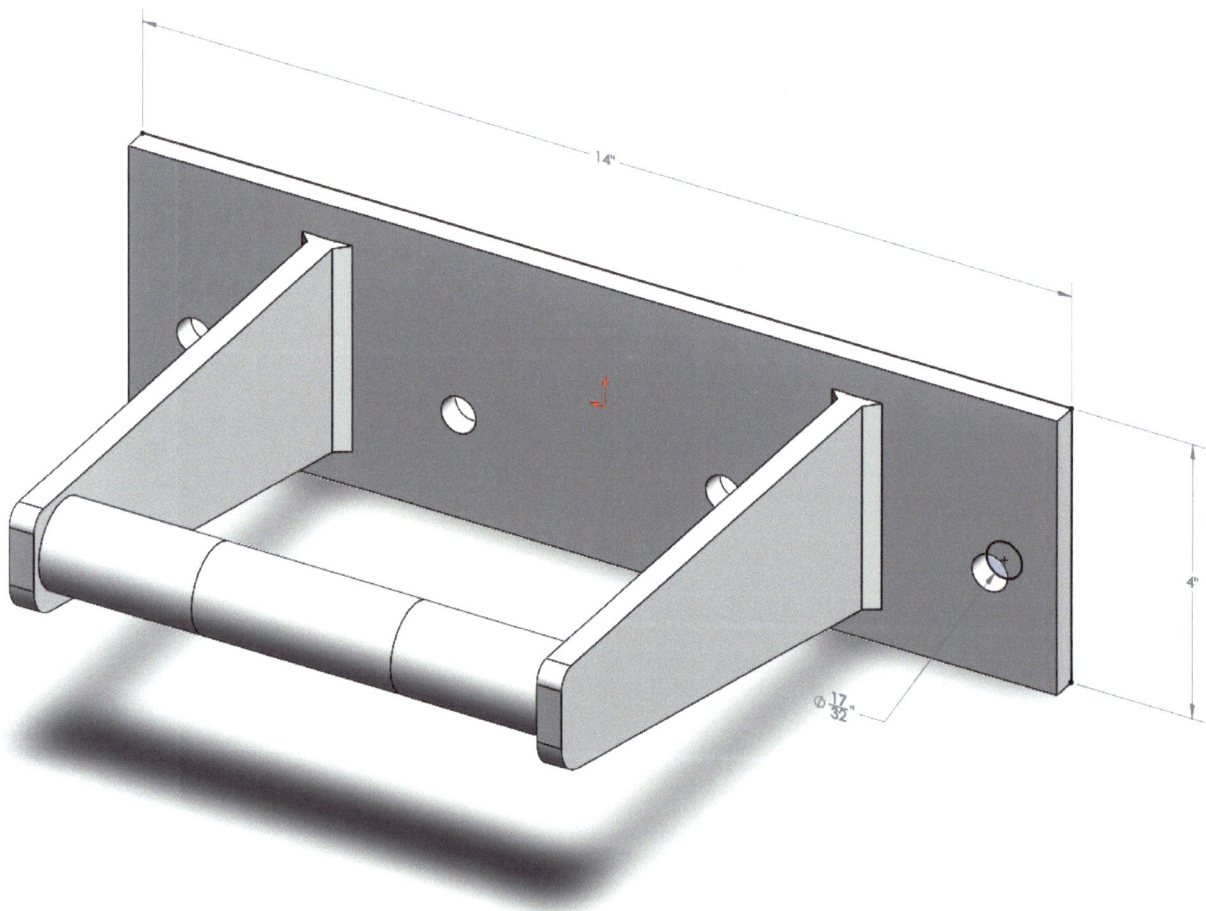

Dimensioned model

Linear Pattern Fixed Face

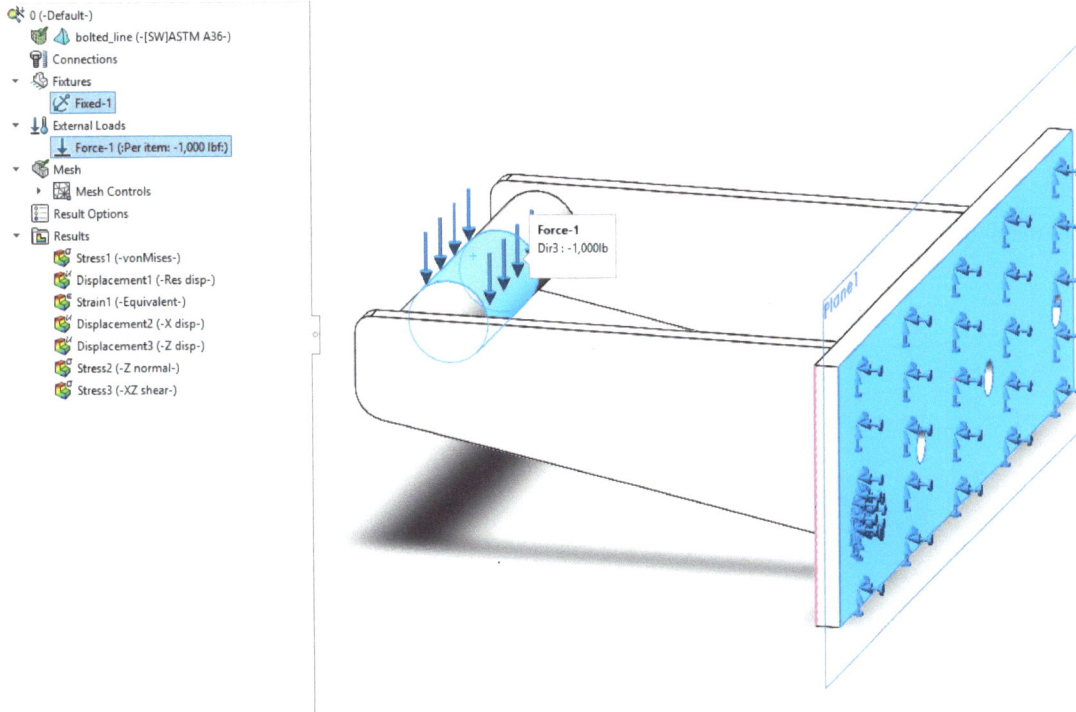

Initial simulation setup

The first simulation setup is very simple. The load is an applied force out on the spool center. The back side of the bolted face is fixed in space.

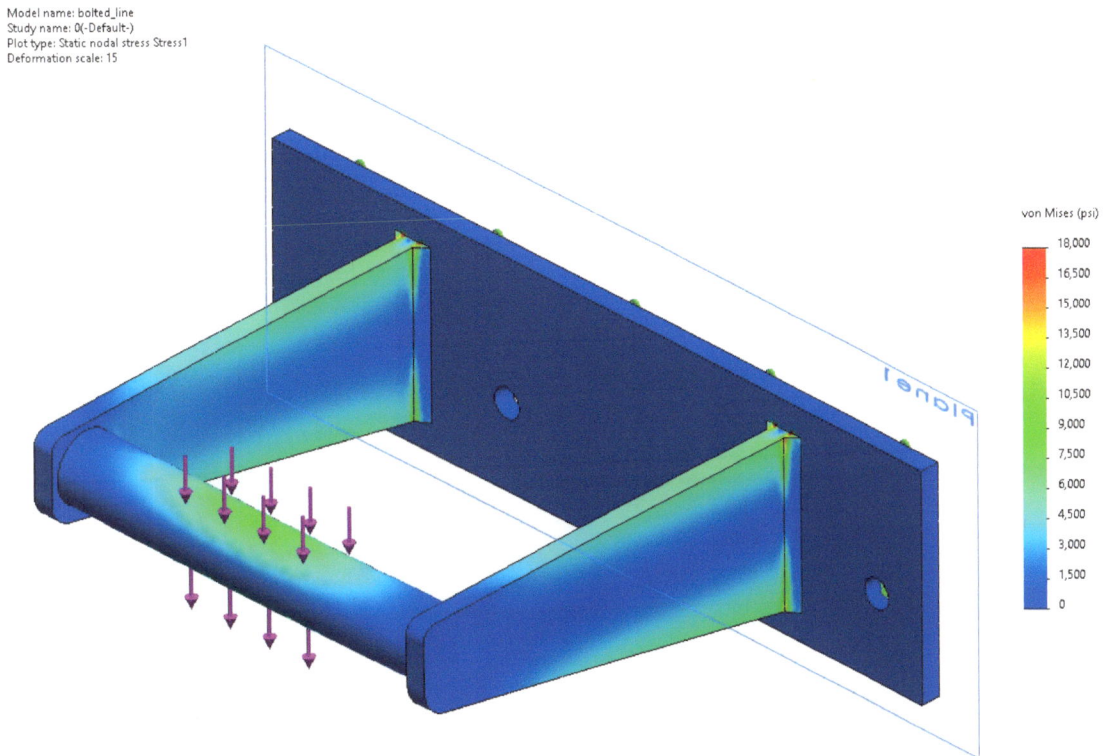

Result with fixed face

The result is predictable. The arms show classic beam bending stress while the fixed plate is practically unstressed at this scale. But with the bolt holes all in a line there's reason to expect the stress on the plate to be more interesting. The top edge might be pulling away from the wall. So as with the previous part we add a contact condition with a virtual wall and foundation bolts.

Linear Pattern Foundation Bolts

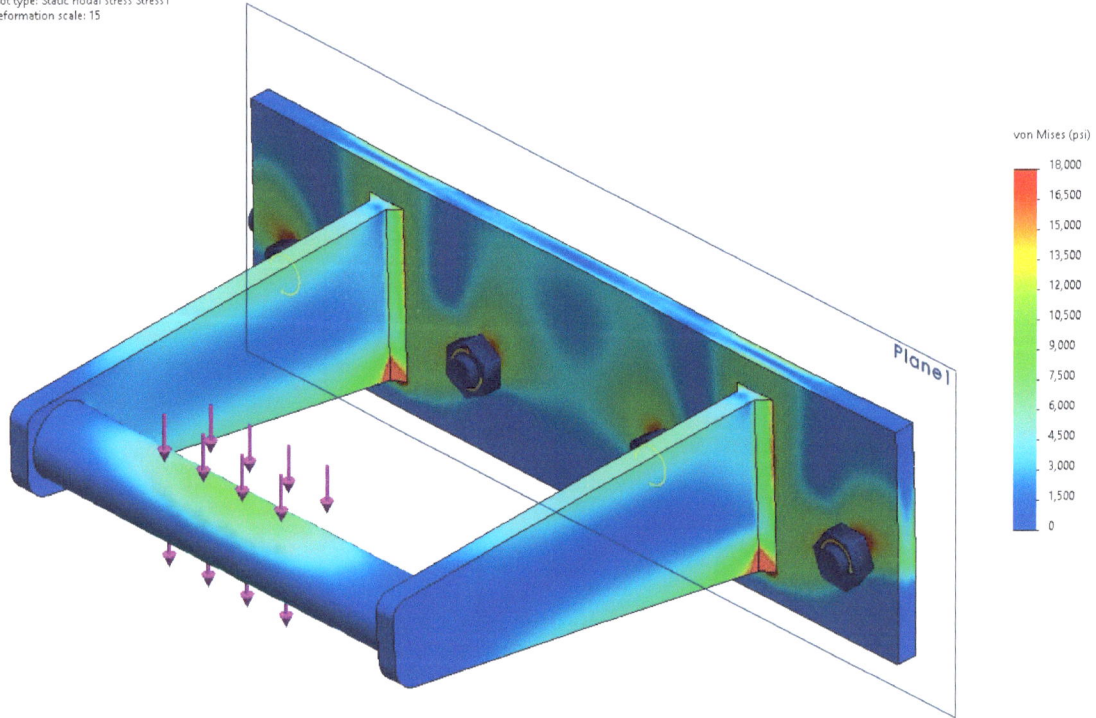

Result with virtual wall and foundation bolts

The difference in response is dramatic. Alternate views show the plate indeed pulling away from the wall (magnified displacement) and a lively stress field on the mating surface.

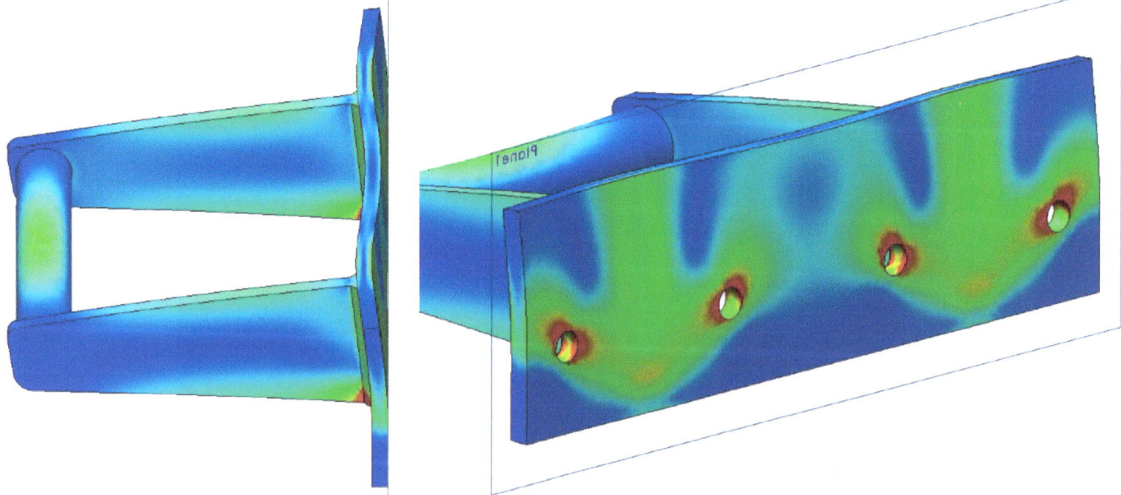

Result with virtual wall and foundation bolts

Linear Pattern, Solid Wall

As in the previous example, we try a solid boundary part instead of a virtual wall. The foundation bolts are replaced with ordinary bolt connectors.

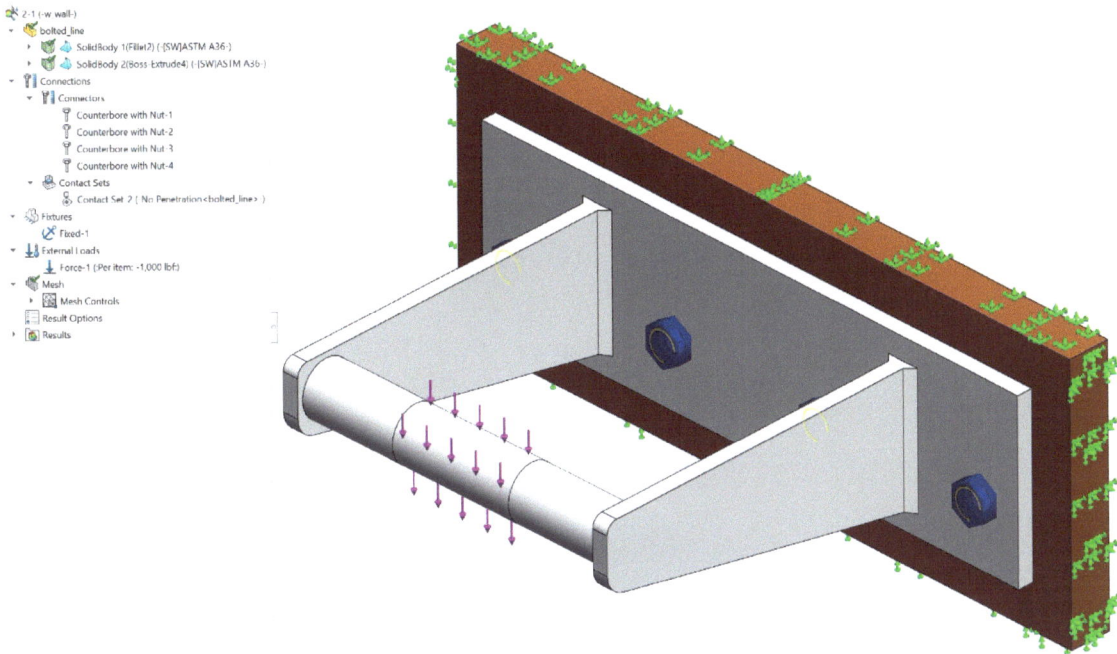

Setup with solid wall block

The bolts here are defined as being "threaded" into nuts on the back side of the solid block. They could have been screwed directly into the solid. This and other bolt properties are not likely to matter on a boundary fixture, but you can experiment easily in your application to see the effects.

A note on meshing

Good results on a surface with no-penetration contact require some attention to the mesh density. Here the contacting face of the solid bock was refined to about double the element size of the plate. The plate itself was refined to get two elements through the thickness, and the weld beads refined further to capture the steeper stress gradients expected. The mesh here is probably more fine than necessary; mesh sensitivity studies are generally recommended.

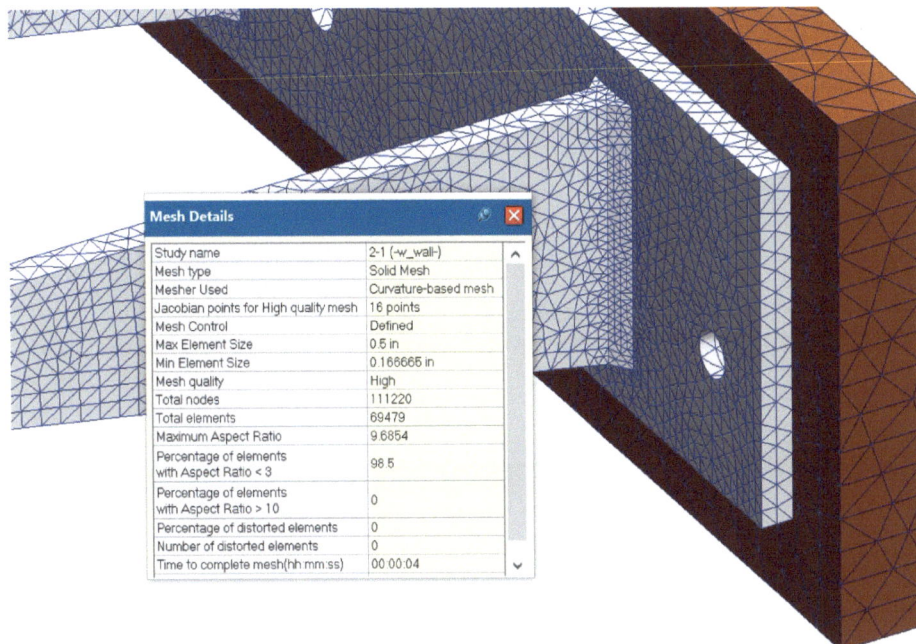

Mesh Details	
Study name	2-1 (-w_wall-)
Mesh type	Solid Mesh
Mesher Used	Curvature-based mesh
Jacobian points for High quality mesh	16 points
Mesh Control	Defined
Max Element Size	0.5 in
Min Element Size	0.166665 in
Mesh quality	High
Total nodes	111220
Total elements	69479
Maximum Aspect Ratio	9.6854
Percentage of elements with Aspect Ratio < 3	98.5
Percentage of elements with Aspect Ratio > 10	0
Percentage of distorted elements	0
Number of distorted elements	0
Time to complete mesh(hh:mm:ss)	00:00:04

Mesh on setup with bolt connectors

Results in the bracket are not noticeably different from the run with the virtual wall and foundation bolts.

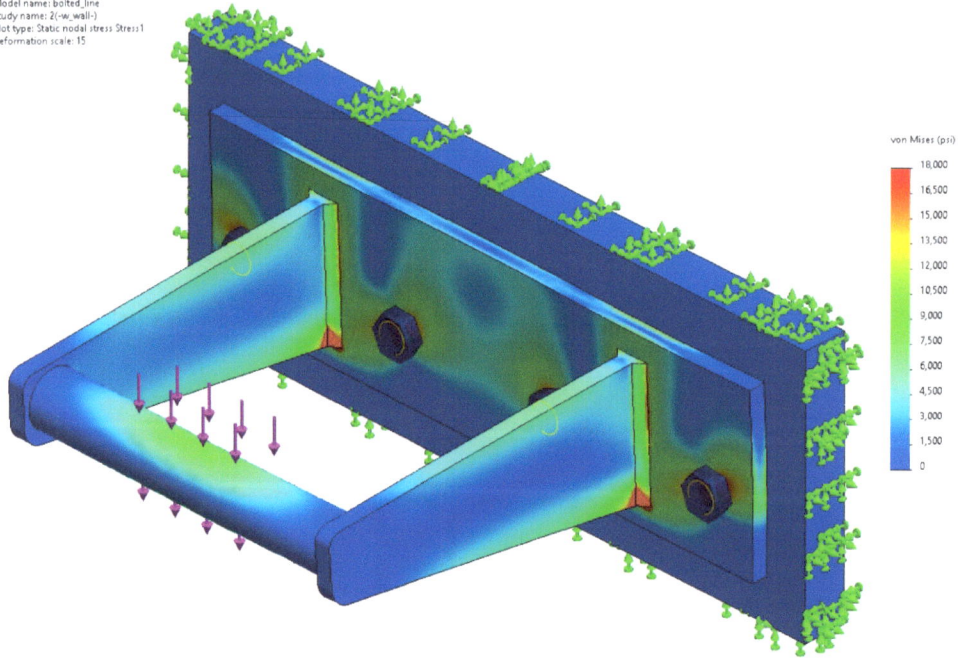

Model name: bolted_line
Study name: 2(-w_wall-)
Plot type: Static nodal stress Stress1
Deformation scale: 15

Results bolted to solid block

Subtle differences can be seen on the back side. And now we can see the stress pattern on the wall surface. The stress scale is reduced for the wall view; unsurprisingly surface stresses on that solid block are low.

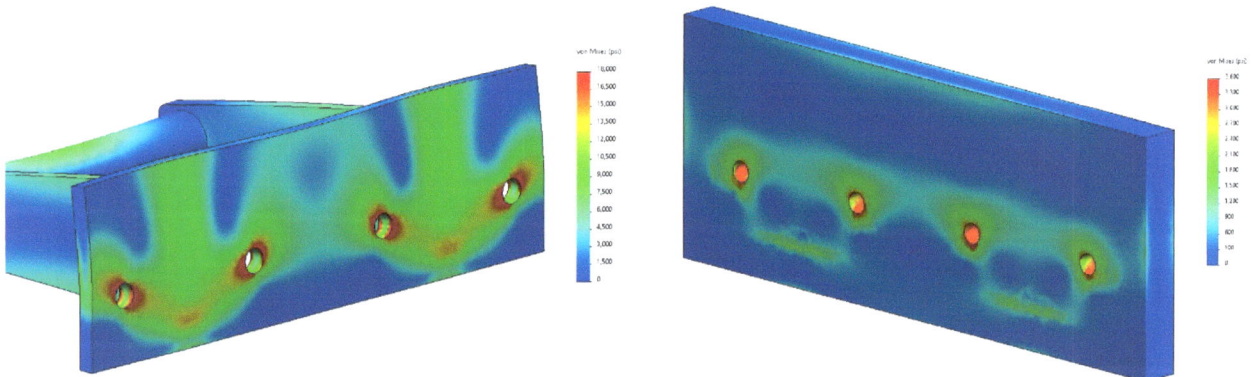

Results on contact faces

Peak stresses are seen near the bolts, so we are curious about what's going on in and around them. The bolt connector definition panel is shown. Far edges are picked (for the bolt-and-nut type). Specific dimensions are input. 12 foot-pounds of installation torque is specified so that a pre-load will be used.

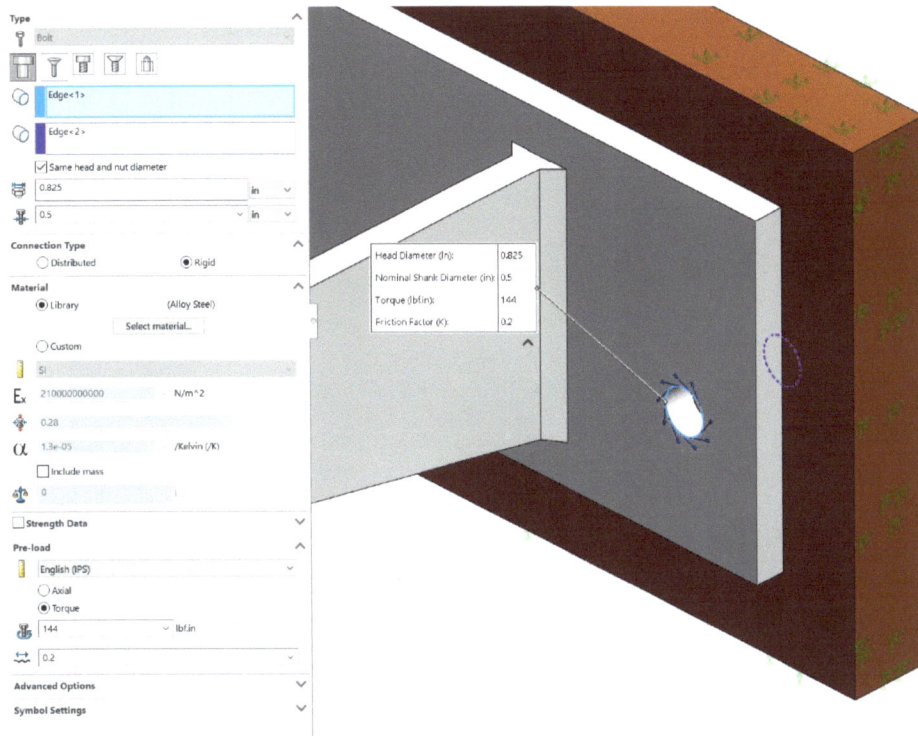

Bolt connector setup panel

The pre-load can be very important in some cases. It will be used along with the input or calculated stiffness of the fastener, and the elasticity of the stack of materials in the grip, to determine the final tension in each fastener. The solver will have to come up with a balance for all of these to get a static solution of the loaded assembly. A large pattern of bolts can drive up the solution time, but they are neglected at the user's peril.

Note the "Connection Type" selection box, which is relatively new. The "Rigid" selection calls up a bolt the way they used to behave. The "Distributed" selection lets some of the nodes touching the bolt move more realistically. The new selection box also appears for the pin connector; we will look into that more when pins come up.

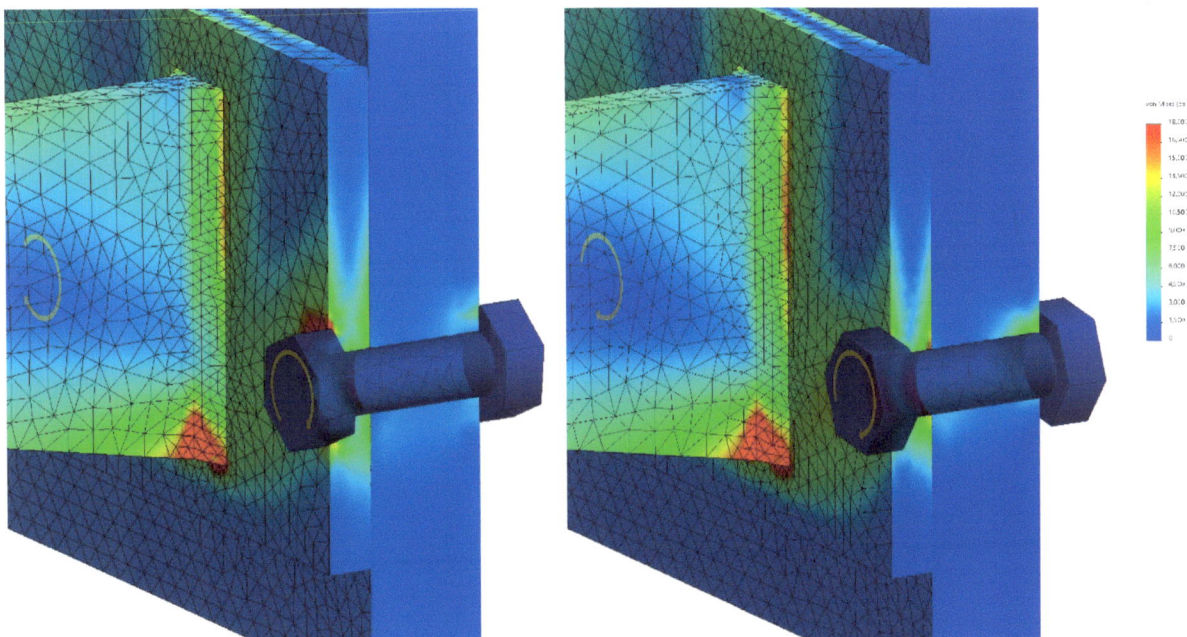

Section through results at near bolt – rigid vs distributed formulation

Literal Solid Bolt

But what if we want to be even more literal and not use a connector for the bolts? There's no reason good bolt and nut models can't be used in this assembly.

Model with solid bolt and nut; pin setup

Four solid bolts and four nuts are mated into the assembly. A "pin" connection is defined from each nut to its mating bolt. Contact conditions are set up between the bolt shanks, bolt heads, nuts, and the respective part faces. Some extra mesh refinement is added to capture response around contact with the fasteners. Does it all work? Sure it does.

The overall result is not obviously different.

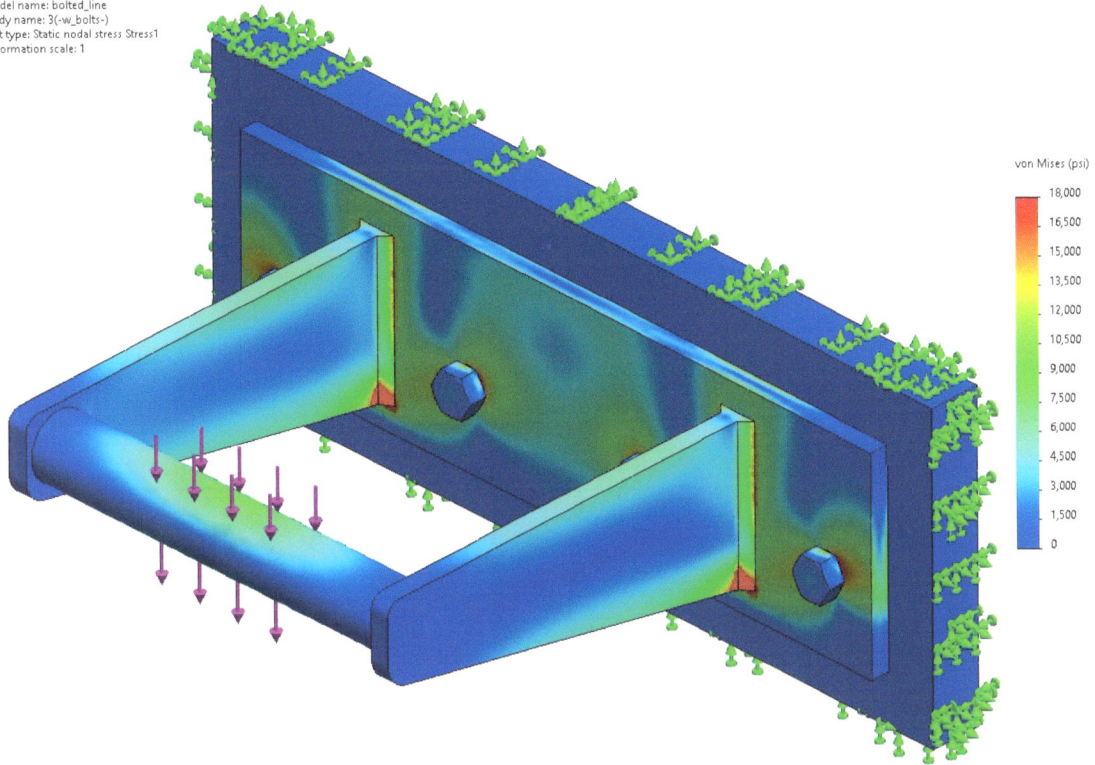

Result with solid bolts

But now we can look directly into the bolts, and we can see contact stresses on the shear plane where the plate wants to push through the bolts [note that the bolt connectors always limit sliding motion].

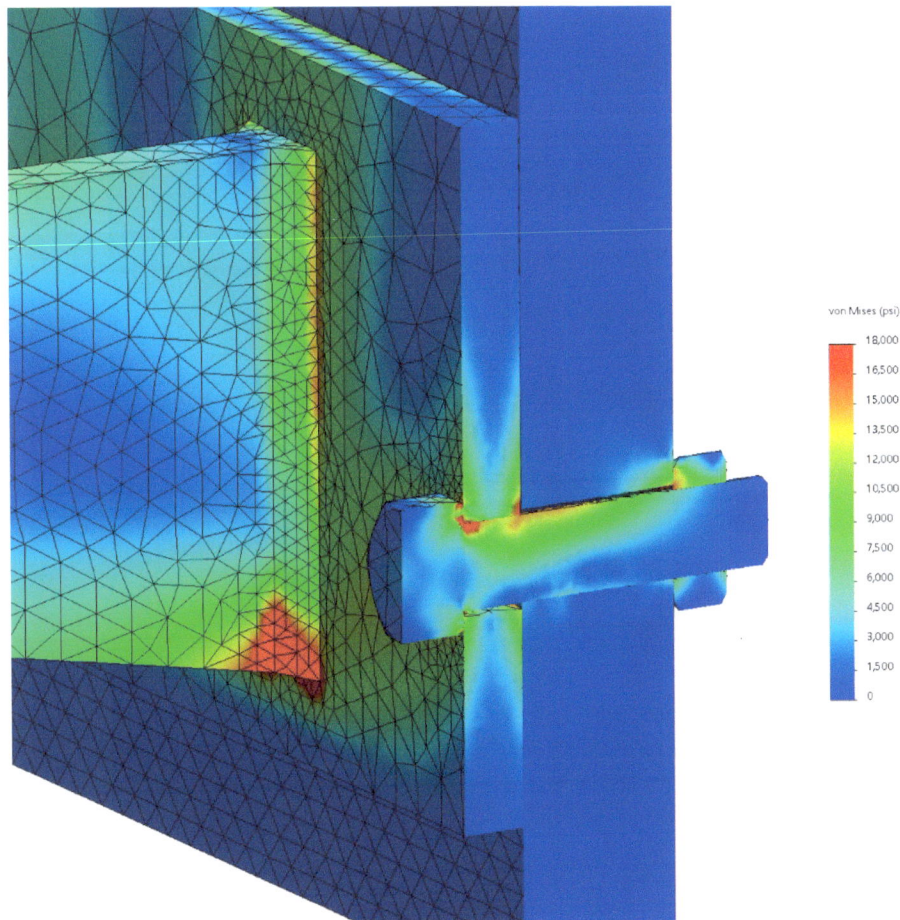

Result section through solid bolt

Stone Lake Analytics

Remember that we lost bolt pre-load in this setup. Axial forces and local compressive stresses will not be quite right. We want to know what else was the cost. Data from the runs is tabulated below. Run time and reference displacement are normalized to the virtual wall study. Bolt force is reported from axial resultant in the nearest bolt connector or contact force under the bolt head for the modeled bolt.

Linear Pattern		elems	run time	disp ref	bolt 1 force
fixed	0	56304	2.2%	22.5%	-
virtual wall	1	57982	100.0%	100.0%	2592
solid wall, rigid bolts	2-1	69479	218.5%	100.4%	2246
solid wall, dist bolts	2-2	69479	255.4%	118.2%	2267
solid bolts, rigid pins	3-1	141577	342.1%	245.8%	1550
solid bolts, dist pins	3-2	141577	437.7%	254.2%	1490

Tabulated study data

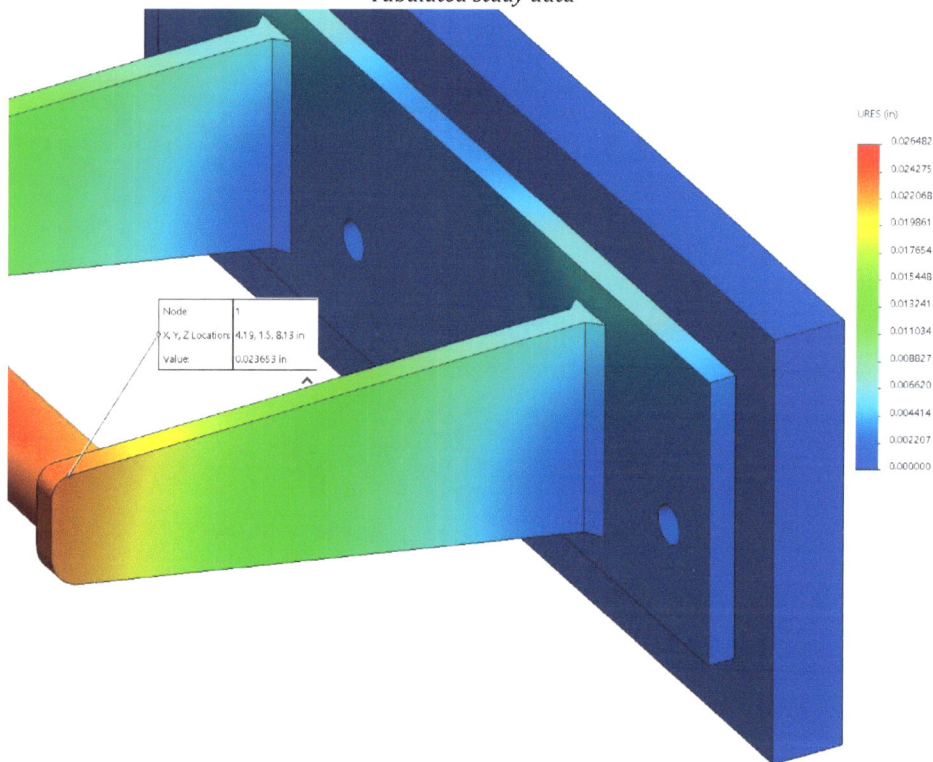

Sample displacement plot with chosen reference point probed

The runs with the modeled solid wall take more than twice as long to run as the virtual wall study. They are also a little more flexible, with lower resulting bolt force. This is to be expected, and it may be more accurate to reality if the real wall material and bolt length are used.

Modeling the fasteners as detailed solids in full contact is of course much harder. The computer must work four times as long. But the result is a detailed picture of what's happening in and around the bolt. Fine modeling of bolts has been found useful in problems where fastener failure was an issue.

Displacements here are elevated largely because of the bracket sliding down into contact with the shank. Final bolt force is lower, but pre-tension was lost in this setup. [We have looked for a way to pre-tension solid fasteners but have not found a clean way to do it that both puts the force in the right spot and works with the actual stiffness of the modeled solid.]

Study 3: A-Arm in Track

Overview
A piston in a slotted track pushes through a short lever on a long lever, which is pinned on one end and resisted by some load on the other.

Objective
Forces in the joining pins and stresses around those joints are needed.

Challenges
Pin connectors do everything needed to connect these parts, but they have some limitations. We must choose between using pins and more expensive methods.

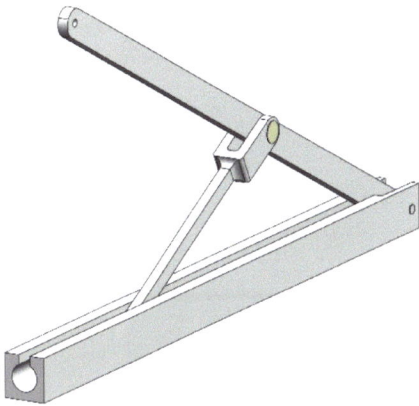

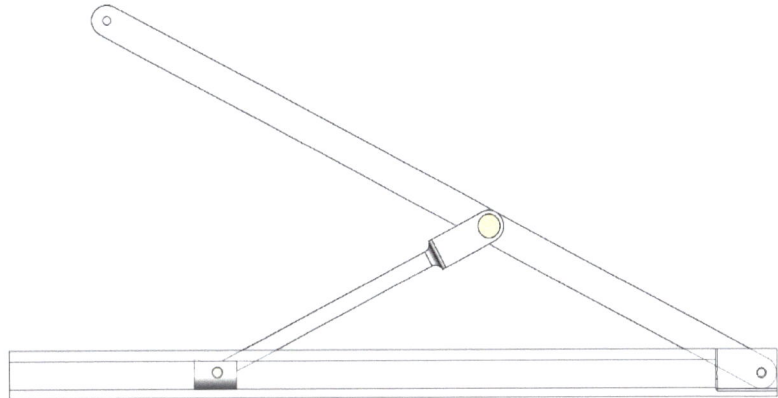

Arm assembly with piston in track

Setup with Pins
The initial setup is shown. The bottom of the track is fixed. The piston is kept from sliding by a "Roller/Slider" fixture which keeps the highlighted face on the same plane. Everything else is a pin connector and the pins are not modeled.

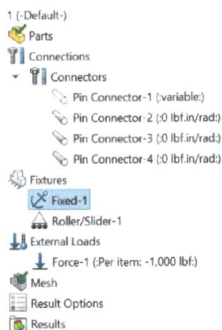

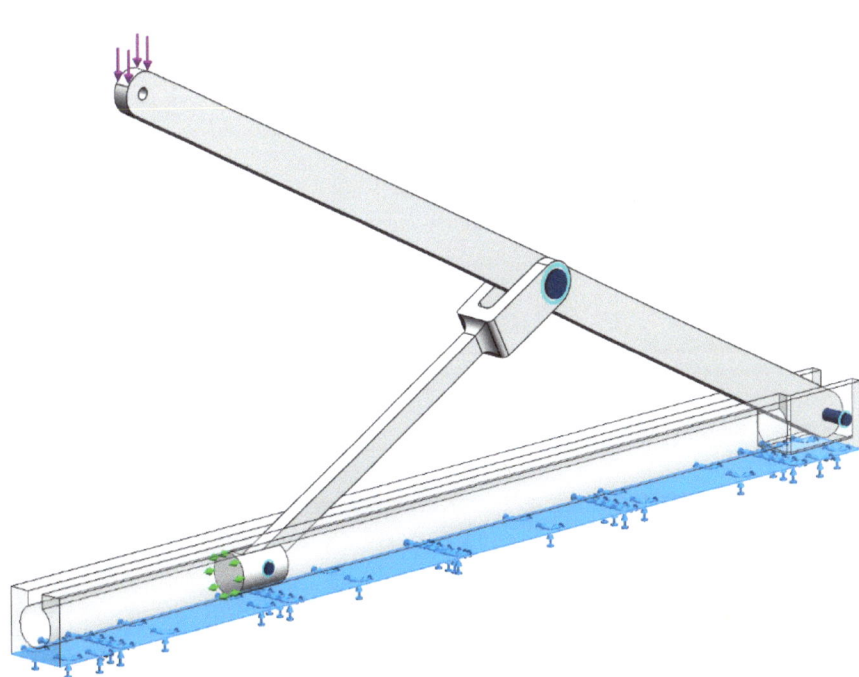

Initial setup with pins

The pin definition on the middle joint is shown on a model cut section.

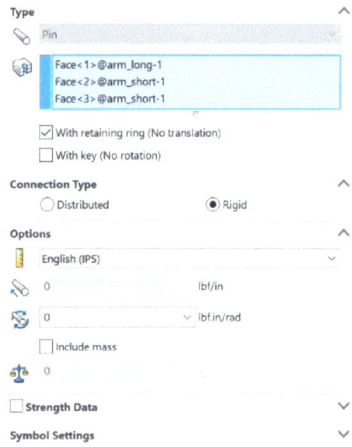

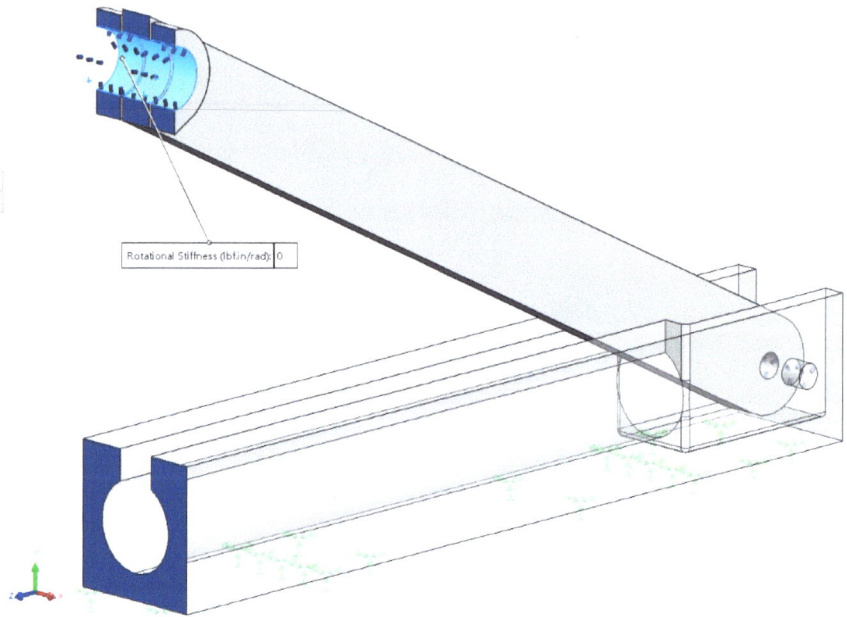

Pin definition on middle joint

All three cylindrical surfaces are picked, one from the long arm and two from the yoke end of the short arm. The simple "Rigid" connection type is chosen, with free rotation but no translation. This will keep all nodes which are meshed on the surface of the three cylinders always on those cylinders. The cylinders can move together in space, and rotate relative to each other, but they will remain perfect cylinders, coaxial and at the same spacing from each other.

The piston is also "pinned" to the track, this time with free rotation. This is a convenient way to handle objects like hydraulic or pneumatic cylinders. Note that Solidworks allows the pin definition even for an incomplete cylinder like in the slotted base.

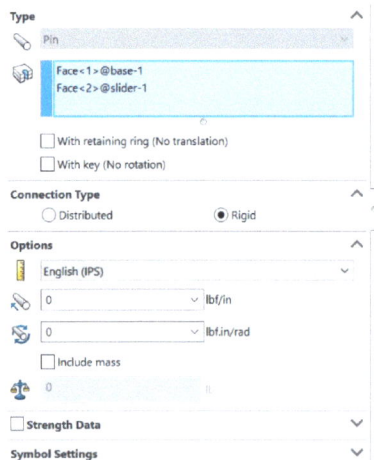

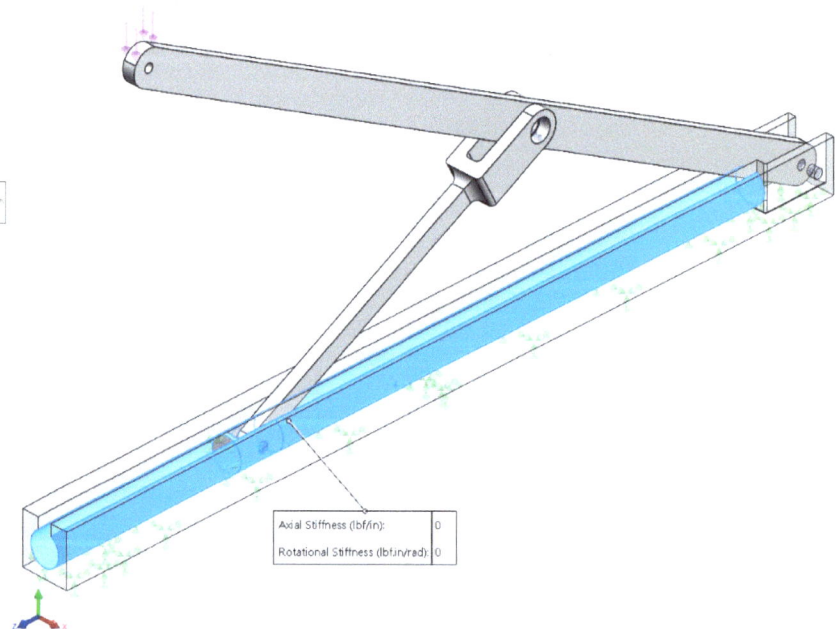

Pin definition between piston and track

The run with this setup takes hardly any time. The results show an expected bending pattern in the long arm

Result with pin connectors

That was easy, but is it right? The best way to check is usually to try it again with every part included, being stubbornly literal about how they all go together. So the connecting pins are all added to the model. Simple double-headed pins are modeled (these might really be headless pins with snap ring retainers on both ends).

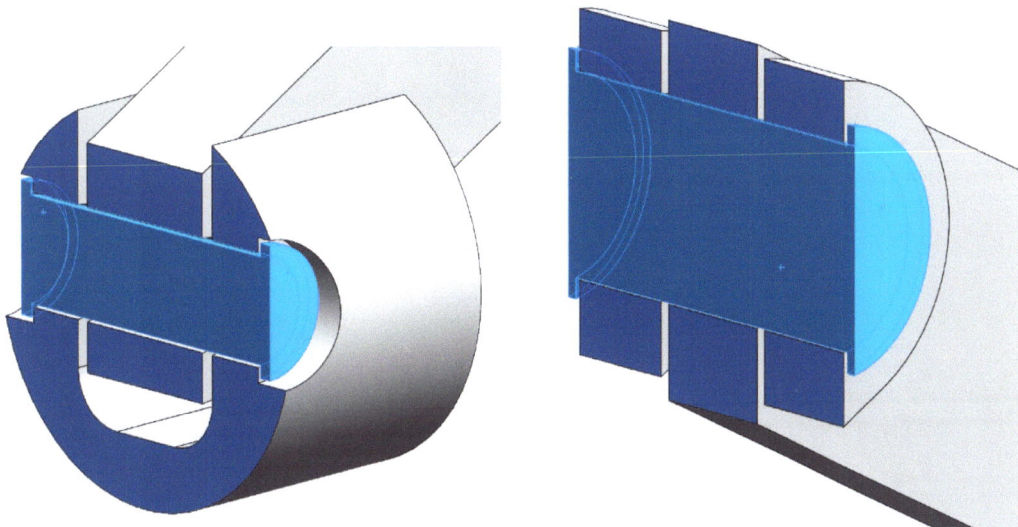

Section views through modeled pins either end of short arm

A solid pin is also modeled at the end of the long arm. Note that there is some clearance between the solid pins and the other components. This slack will be taken up in the simulation result. For a good result and fast solve it is best to keep these clearances small.

Setup with Global Contact

We must tell the program what to do with the new parts. The easiest thing to do is to define a "Global Contact" condition.

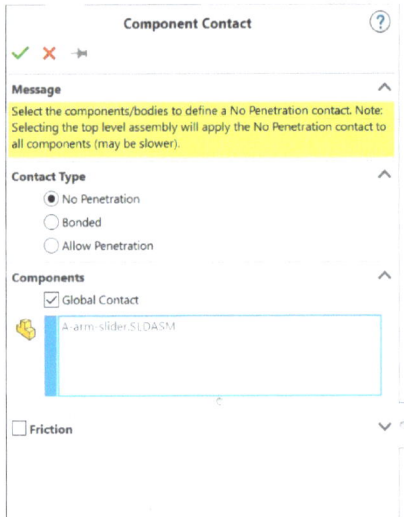

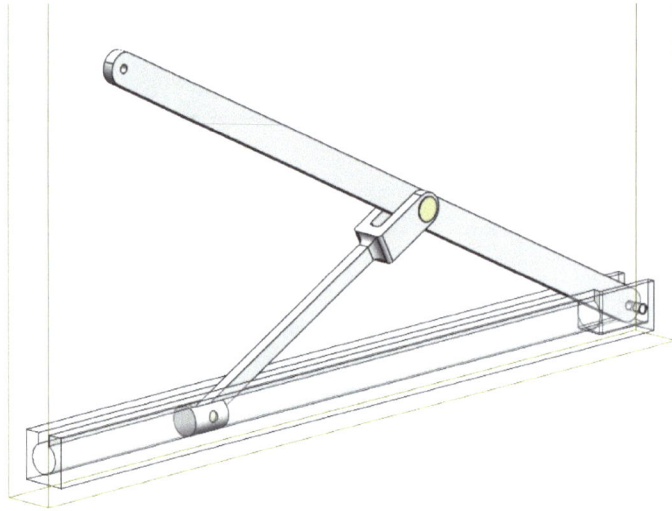

Global contact condition setup

The solver will consider every face of every component possibly in contact with every other face of every other component. For a larger assembly this can take some serious calculation time to sort out (and in some cases one will run into internal program limits) but for this small assembly it's no problem.

Result with global contact

The result looks similar to the first, but a close look around the middle joint shows a strong difference. Results are plotted on a section just inside the near yoke arm.

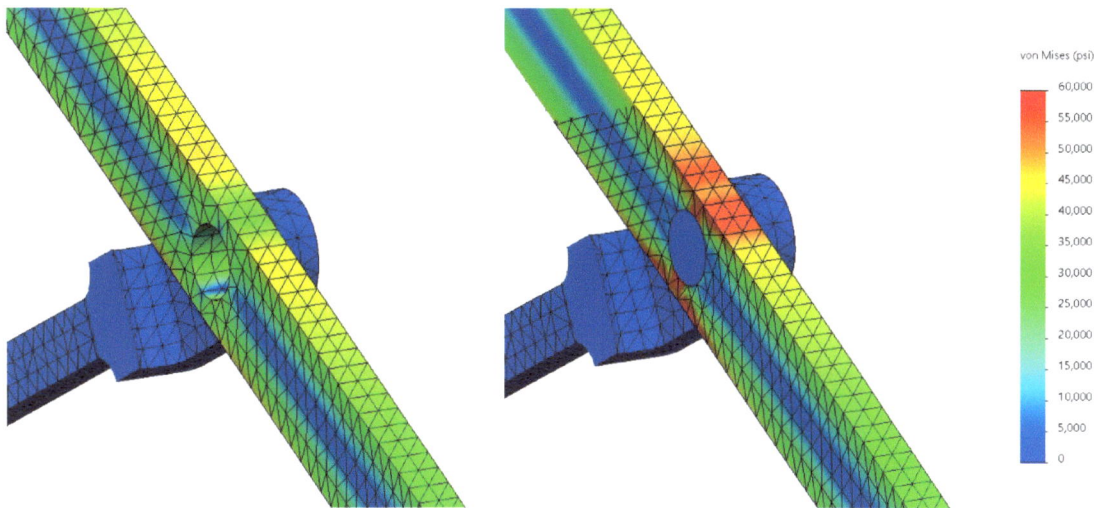

Result with pin connectors vs with global contact

Stresses in the long arm are much higher around the modeled pin in contact. Remember that the synthetic rigid pin connector is 'sticky' – it holds on to the nodes around it, keeping them in a perfect cylinder. Strains for some distance around the pin will be artificially low.

Including all the real geometry and putting all the parts into no-penetration contact is usually the best way to capture the real behavior of mechanical parts. We did it here with a single pick of a global condition, but there are reasons that we will want to define contact manually.

Manual Contact Sets
This is the tedious part.

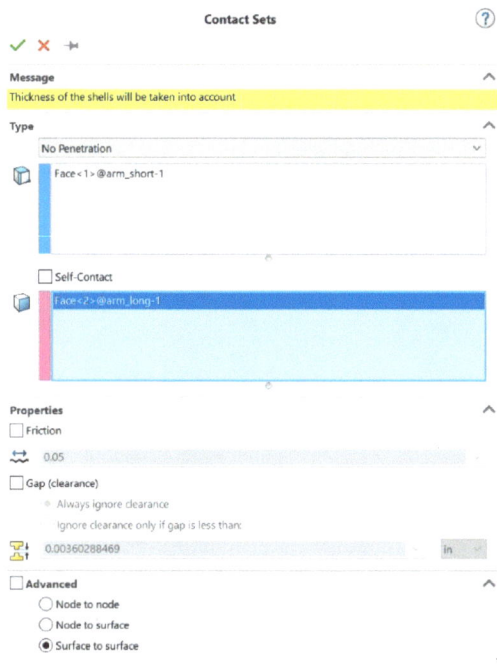

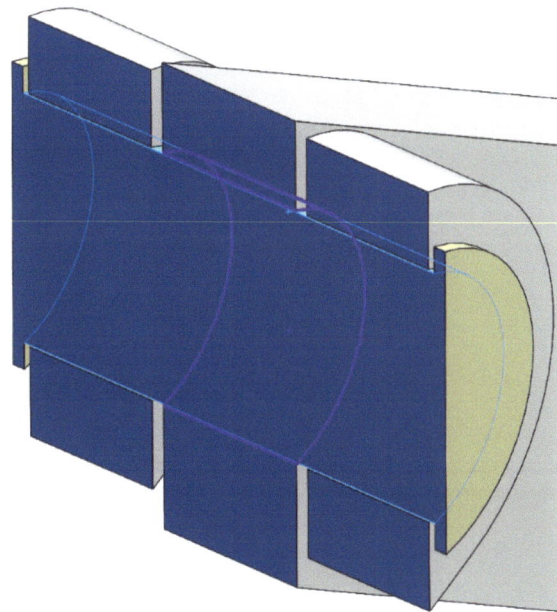

Definition of contact pair

Here one cylinder, from the middle of the long arm, is being put in contact with the body of the pin. Other pairs of surfaces are connected similarly. Working on section views or in wireframe and using the "Select Other" function are helpful for this task. [Remember that a main strength of Solidworks Simulation is access to the whole solid modeling front end and user interface!]

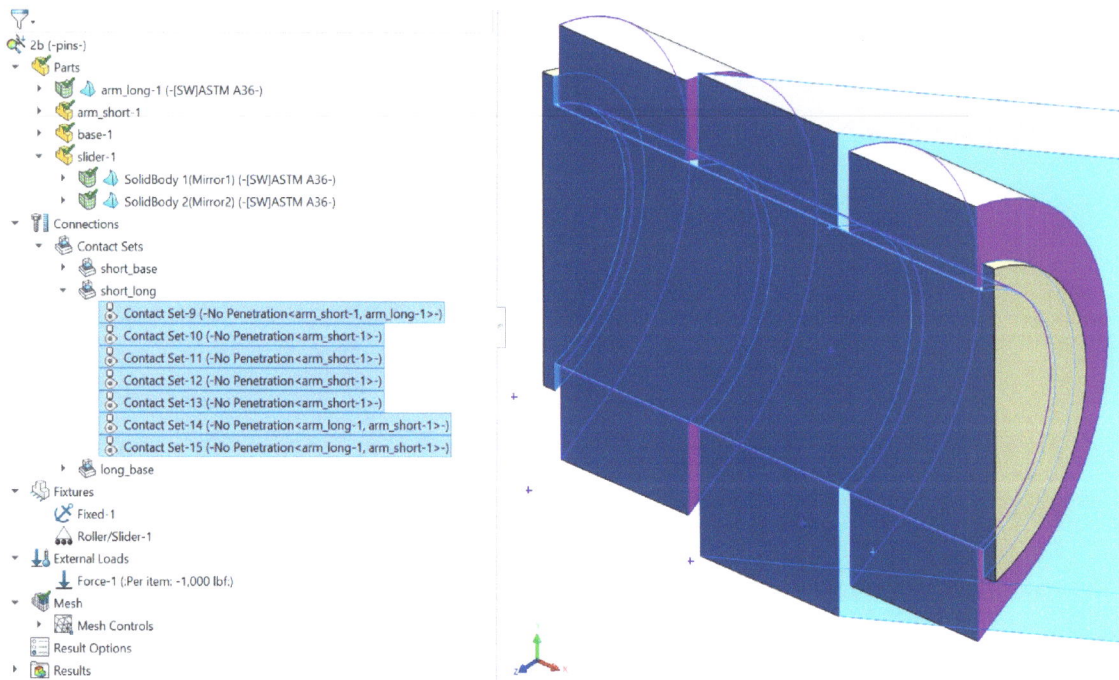

Contact set for one joint

The contact sets for this one joint are listed. Note that contacts are added between flat faces, as otherwise free motion would be allowed in the axial (X) direction. The contact sets are organized into folders as each group is made – this is highly recommended.

In this case manual definition of contact sets does save some solve time, but it is a matter of seconds, and the result is identical.

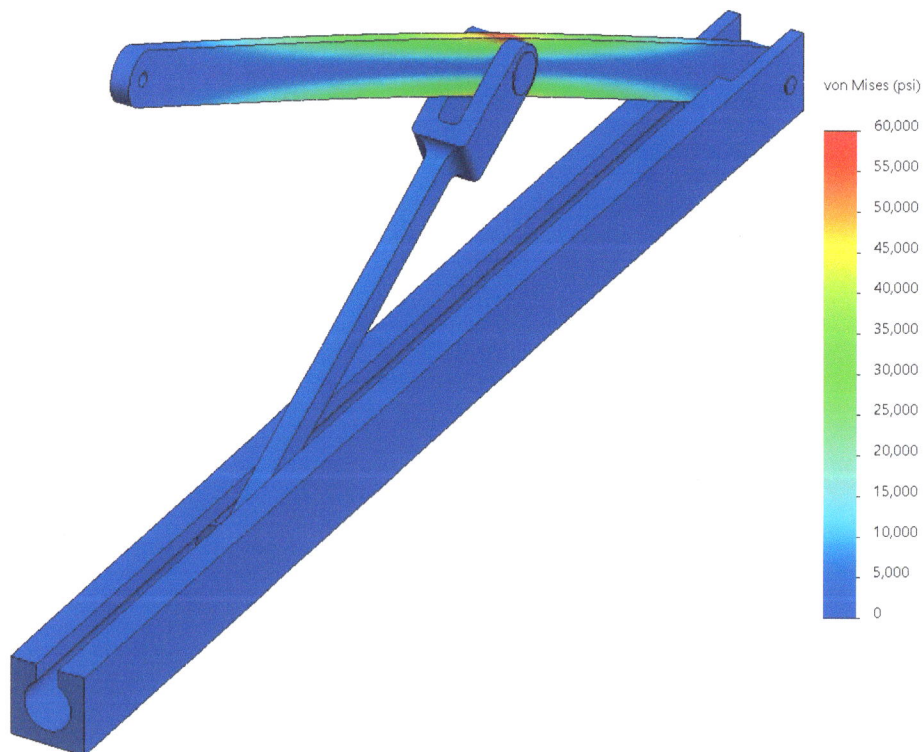

Manual contact sets will be very important to later examples.

For serviceability it might be required that the middle joint be greaseable. So a hole is drilled through to the center bore of the arm, with a small counterbore where a grease fitting can be installed.

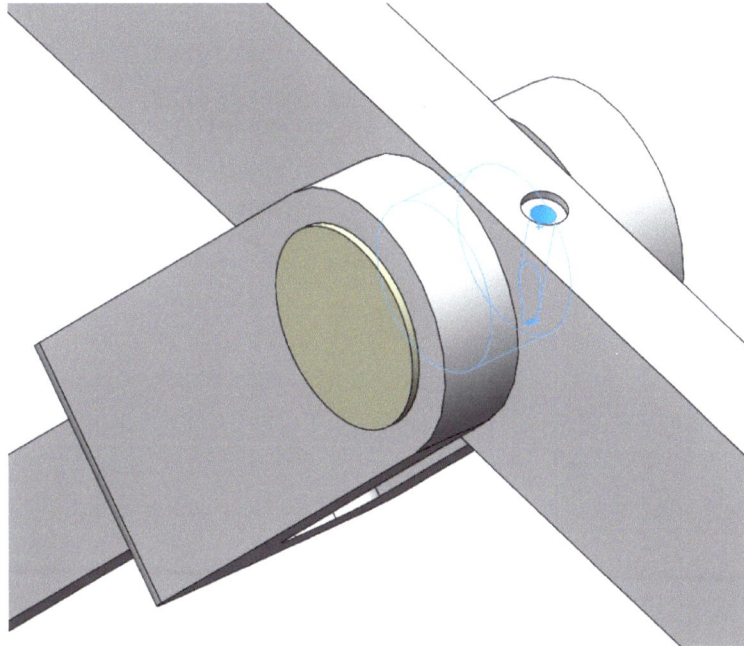

Model with intersecting grease hole

Some mesh refinement is added around the small and large holes and the studies run again. Results on a central cut section (with mesh lines) are shown.

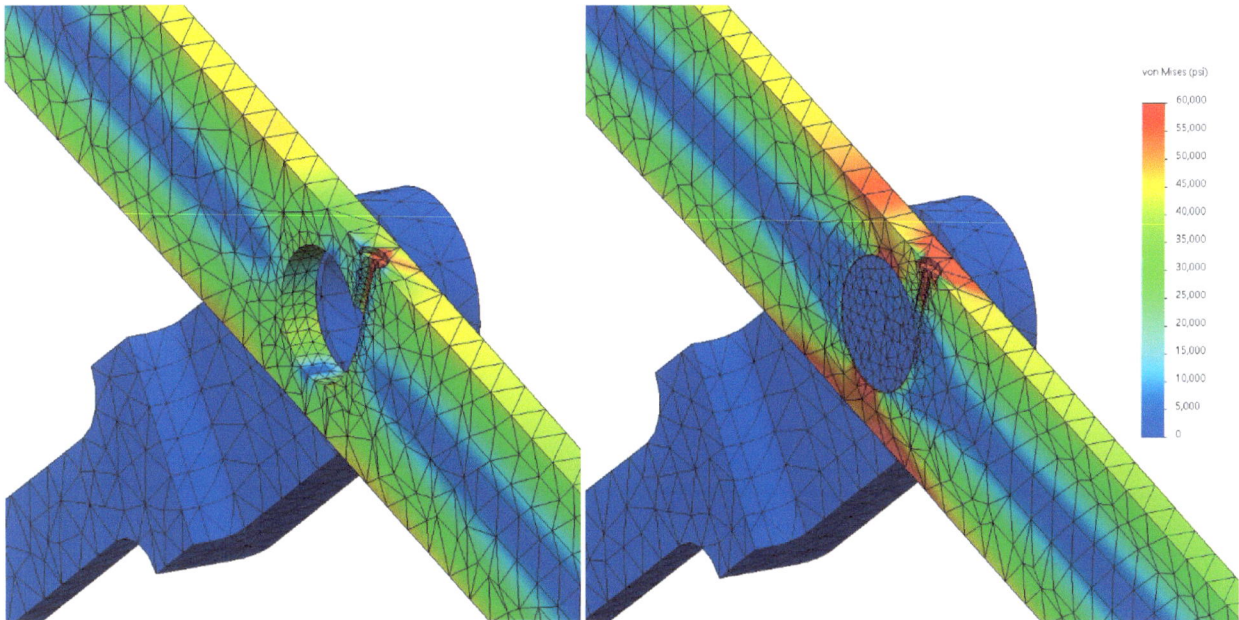

Result through small hole, rigid pin vs full contact

The result around the synthetic pin connector does show a stress riser from the small hole. But the stress field all around the modeled pin is higher and the small hole much more of a problem.

Arm and slider			elem	run time	max disp
1	pins not modeled	rigid pin conns	56983	100%	100%
2a	pins, headed	golobal no-pen contact	60308	2467%	135%
2b	pins, headed	manual contact	60308	2278%	135%
1cr	pins not modeled	rigid pin conns	61234	111%	100%
2cr	pins, headed	manual contact	70471	3456%	136%

Summary of runs

Tabulation of the studies shows that the full contact studies took much longer; they also gave much better results. Displacement in the contact studies is higher, owing to both slack take-up and compliance in the solid pins.

Study 4: A-Arm with Roller Track

Overview

The same arm arrangement as in the previous study is used but the piston is replaced with a rolling axle and the slotted track now has a square cross section. Another loose piece is used to take the external restraint. A real assembly here would include rollers with many internal parts and a method of attaching them to a shaft. It is assumed here that all of that is done effectively and that the rollers can be considered merged with a solid shaft.

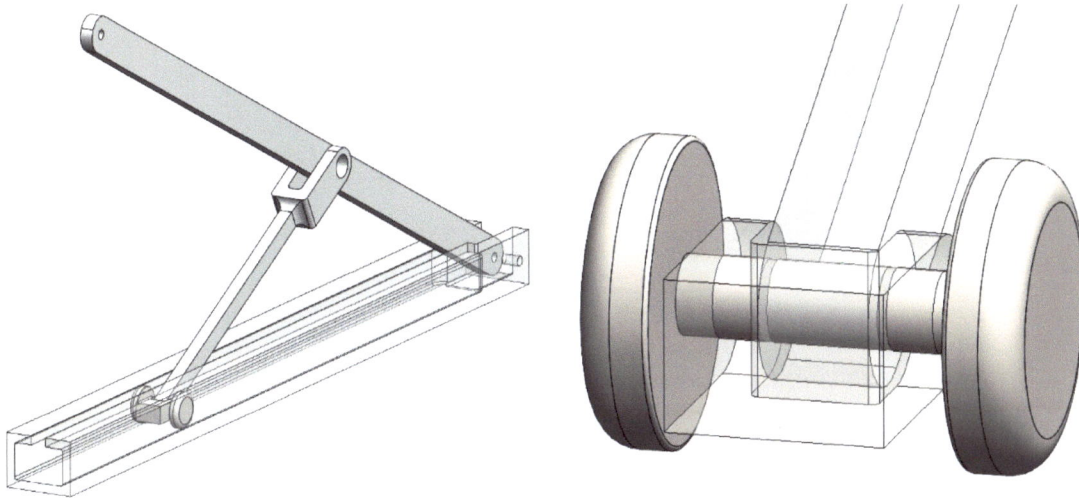

Modified assembly

Objective

We want to know forces in the rollers, including side loading.

Challenges

As modeled small surfaces are in contact with much larger surfaces, and some surfaces can contact multiple others. Reporting of resultant forces in Solidworks is sometimes not helpful in these situations.

Setup

The roller body is modeled so that clearances are small to all other parts. This is recommended for all contact studies, at least at the start, for fast clean results. Later one can investigate the effects of a poorly shimmed build or larger tolerance stacks.

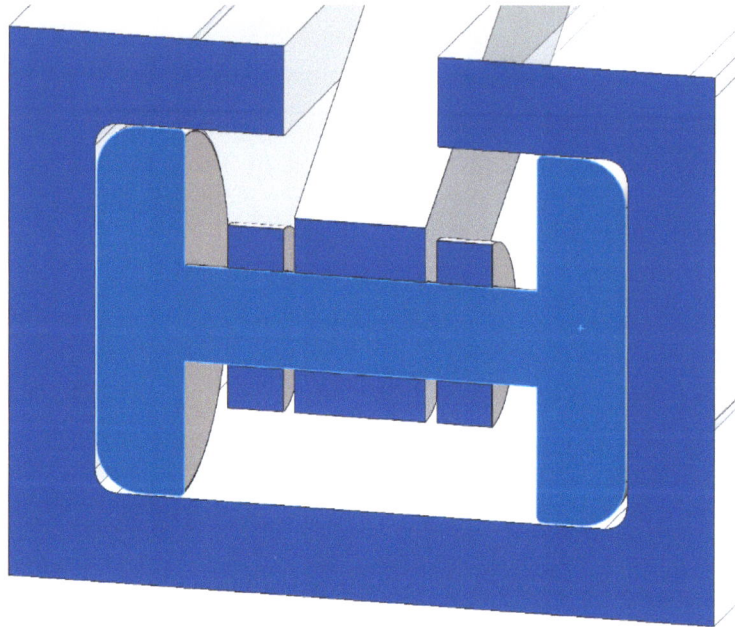

Section through roller assembly

Study setup is as before. The other joints remain simple rigid pins as we only need their articulation, not detailed stress results. A global no-penetration contact set is used, which should cover whatever the rollers do inside the track, without us having to spell it out.

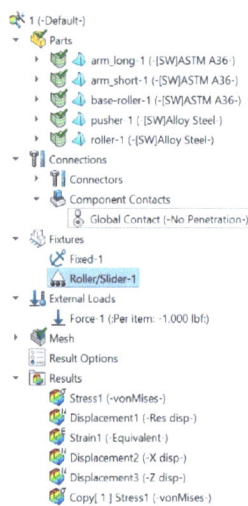

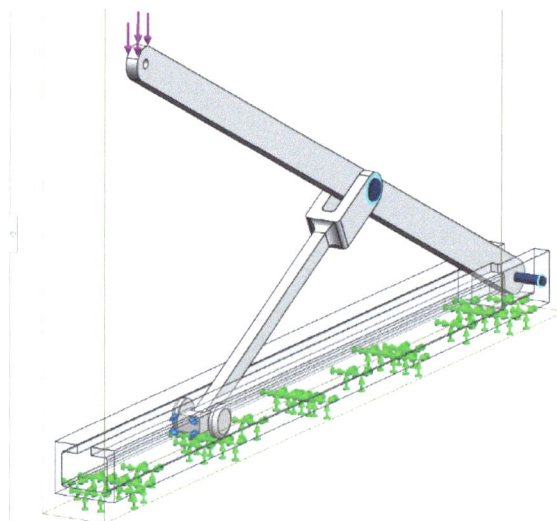

Study setup

For the roller axle we switch to the "distributed" pin connector, which will allow bending along the axle. Note that now the male portion of the pin joint is in the model. Previously in pin joints the shaft was omitted and only cylindrical holes were used. The Solidworks Simulation pin joint gives us the flexibility to use it either way.

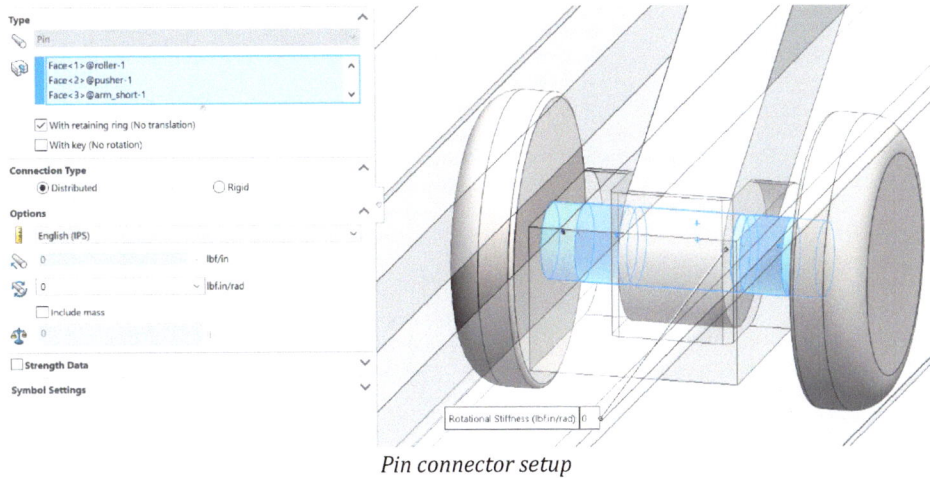

Pin connector setup

Note that we have left a free degree of motion here. The roller is technically free to rotate. The solvers in Solidworks Simulation are usually forgiving of such a lapse if there are no forces in the study to agitate the partially loose part. [The direct solvers may be better for this than the iterative solvers, but we have not experimented on this point.]

The overall result is the same as the previous example. We switch to a lower stress scale to get some detail around the new roller.

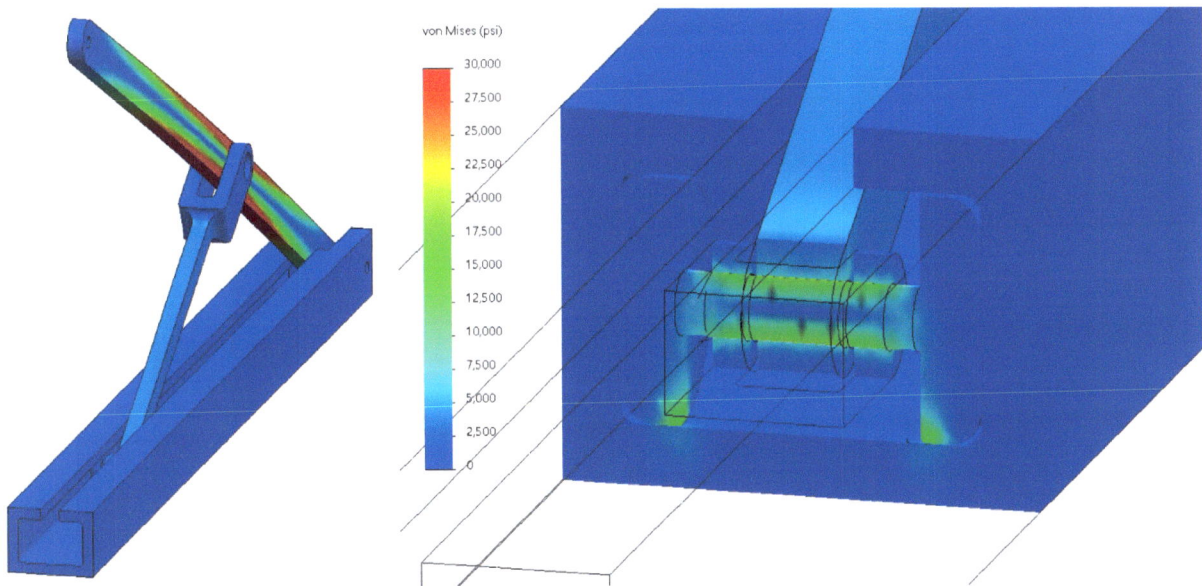

Overall stress result, stress on section through roller

We want to know the forces on the rollers so we pick "Contact/Friction Force" under "List Result Force". All the outer faces of the right roller are picked – we don't know which faces would have been found in contact.

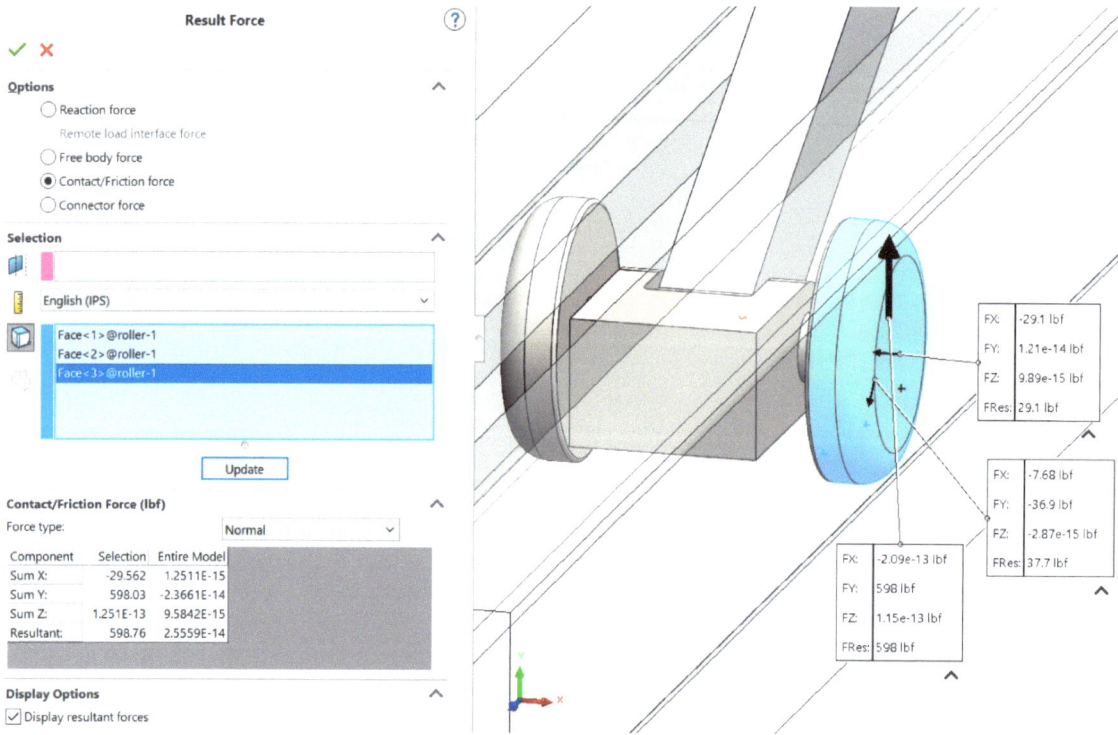

Contact forces with global contact condition

The large arrow is the vertical force which we expect is from the bottom tangent point on the cylindrical face of the roller. But since that face wraps around the whole roller the force vector is displayed somewhere in the middle. And if there was some interaction with the top of the track, we can't know.

Lateral (X) forces are seen from both the flat face and the outside radius of the roller. This is expected as the tangency between the two does approach the side face of the track. But for simple tabulation of the forces, such as for comparison to catalog component ratings, it might be handy to have just one clean lateral contact.

Splitting Faces for Contact

We edit the model to split the contacting faces for both convenience and efficiency. Here the contact setup is shown with faces picked for the left roller.

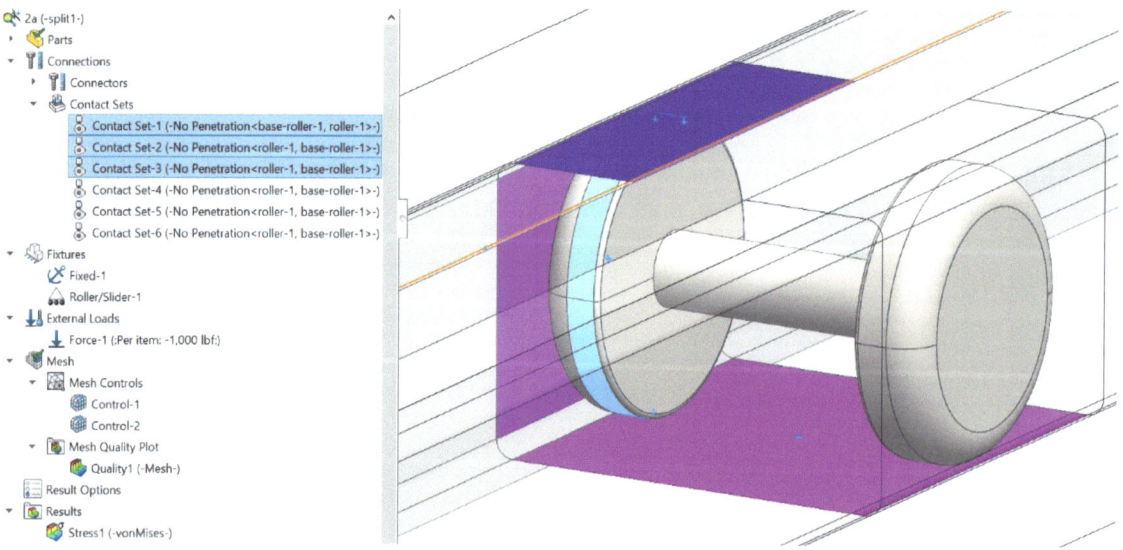

Contact setup on split faces

Solidworks Simulation for Real Machines

The cylindrical face of each roller is split into top and bottom halves. Each half is put into contact only with one face of the track. The inside faces of the track are also split so that only a small area around the roller is considered. This allows only that part of the track to be considered in contact analysis, speeding up the solve, and allows us to apply local mesh refinement without refining the entire track.

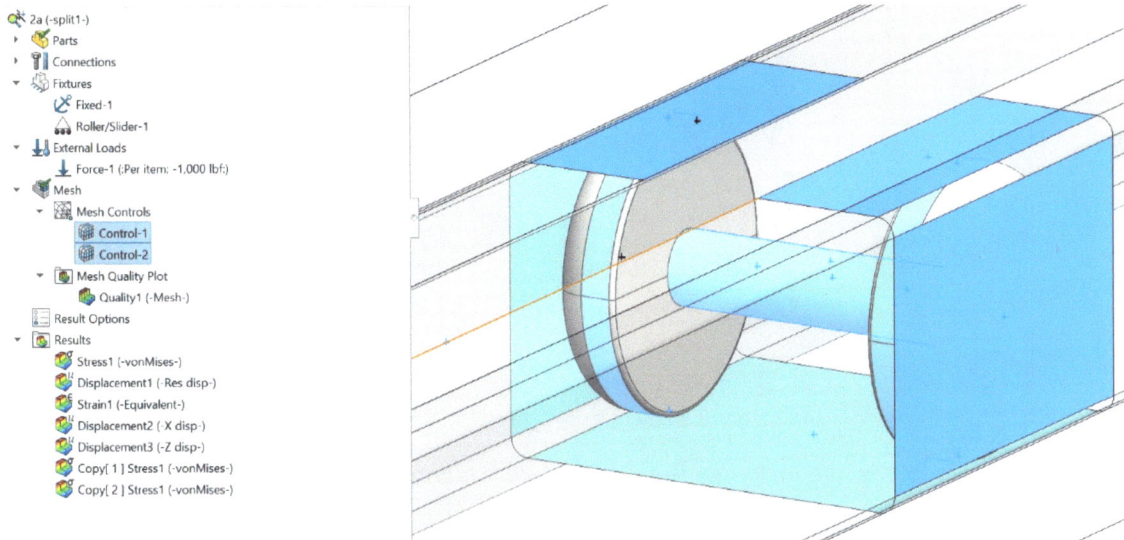

Selective mesh refinement on rollers and track

We can expect a little more flexibility in the model with the refined mesh and more precision in the contact results. A plot with magnified displacement shows the rollers splaying outward. Stress plots on the rollers clearly show the contact points.

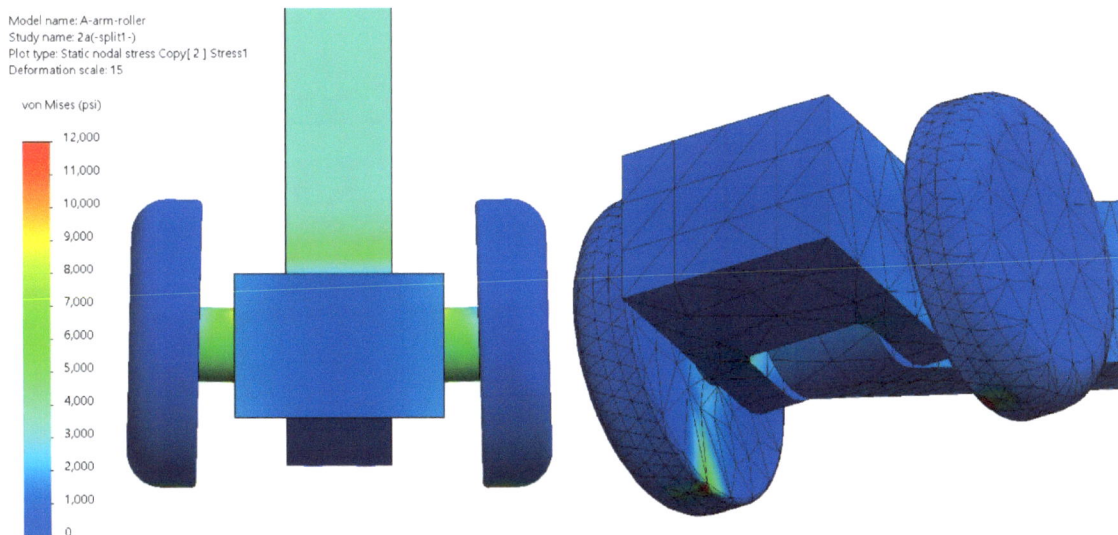

Stress plot with magnified displacement (track hidden)

Since we were able to refine the mesh locally we can see sharper stress peaks and more detail in the stress field.

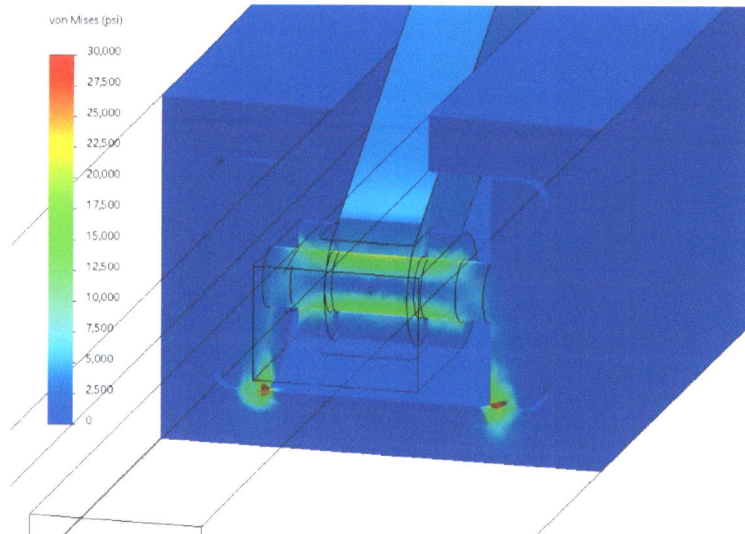

Stress result on section through roller, split model

When probing contact forces, we pick only the surfaces which were put into contact.

Force probes on split surfaces

Hertzian contact stress

Nota Bena: Finite element methods are capable of capturing accurate stresses at line or point contacts of balls and rollers. You will *not* make these predictions in production analyses of large assemblies. The following shot is from a Solidworks knowledge base example. The mesh density here is on the order of what is needed to approximate the exact theoretical stress field near a point contact of elastic materials.

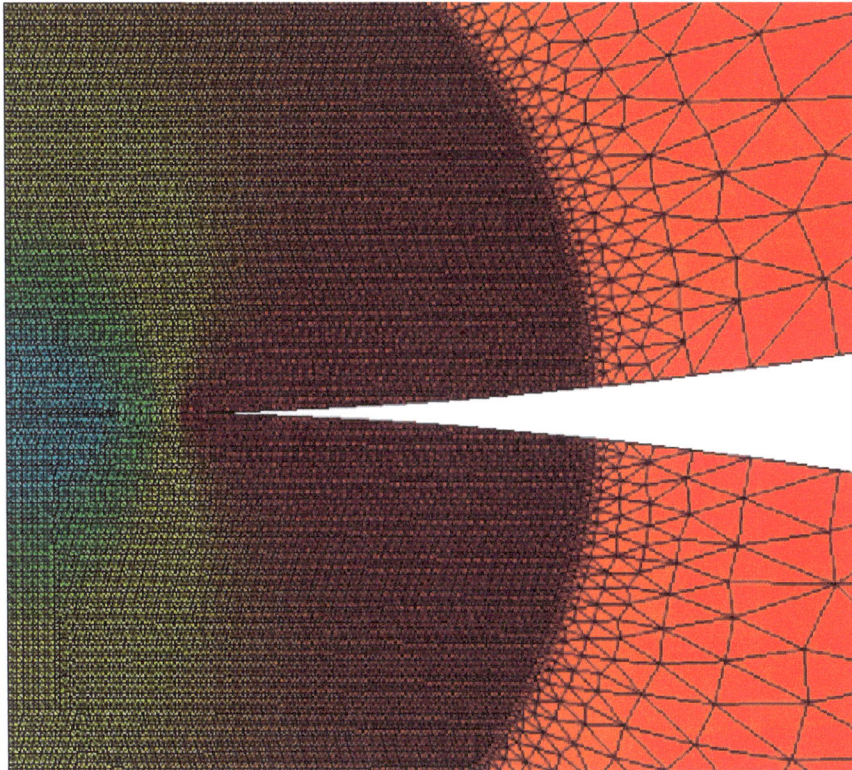

Good stress result for Hertzian contact

The reason we get force data from a model is to use it for rating connecting hardware. With forces in hand one can use a simple Hertzian calculator tool to get peak stress and total deformation predictions.

Curvature-based Mesh

So far we have used the "standard" mesher. Now we will try the "curvature-based" mesher.

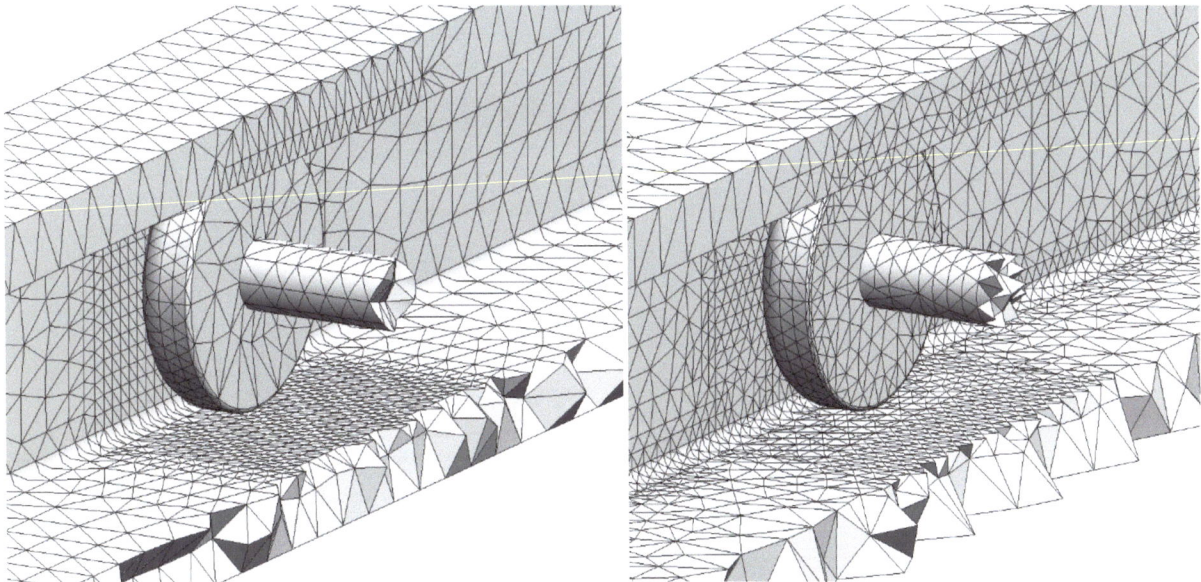

Mesh sections from standard and curvature-based meshers

The curvature-based mesher gives some automatic refinement through radii, whether desired or not. In many cases this conveniently refines the mesh just where stress gradients are found. Here it refines through the convex radii on the roller, which is probably good. It also refines the entire lengths of the concave radii inside the track, which is probably unnecessary.

Stress results are compared.

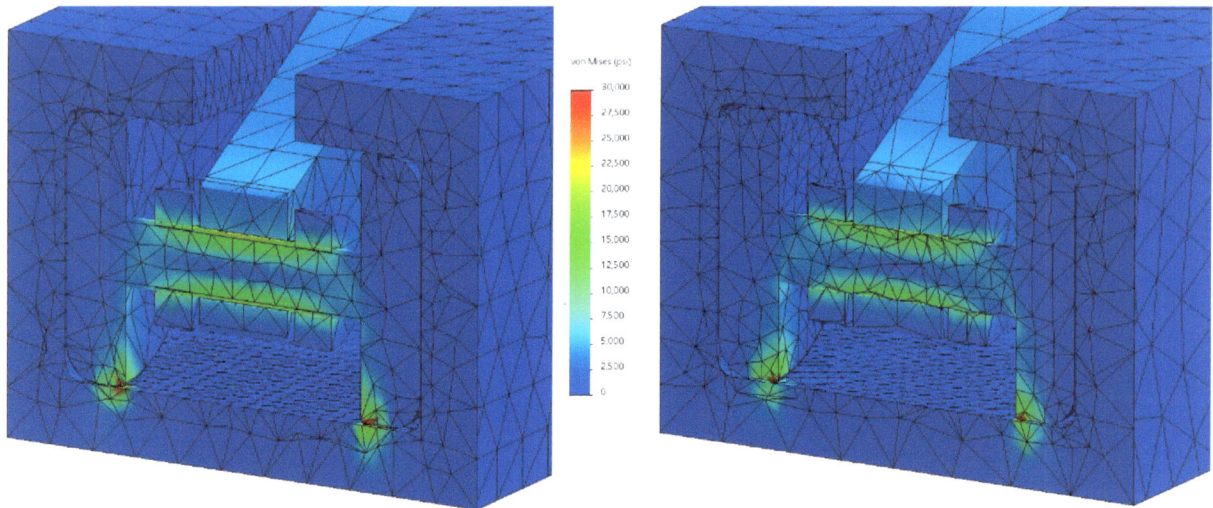

Result through roller, standard refined mesh vs curvature-based mesh

There are subtle differences in the stress response.

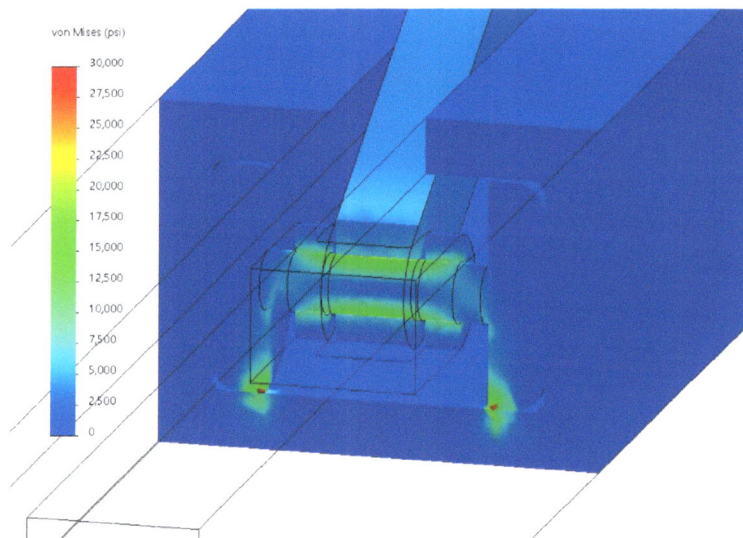

Result through roller, curvature-based mesh

The lateral force increases a bit more, likely owing to additional precision in the contact solve and flexibility in the roller-axle.

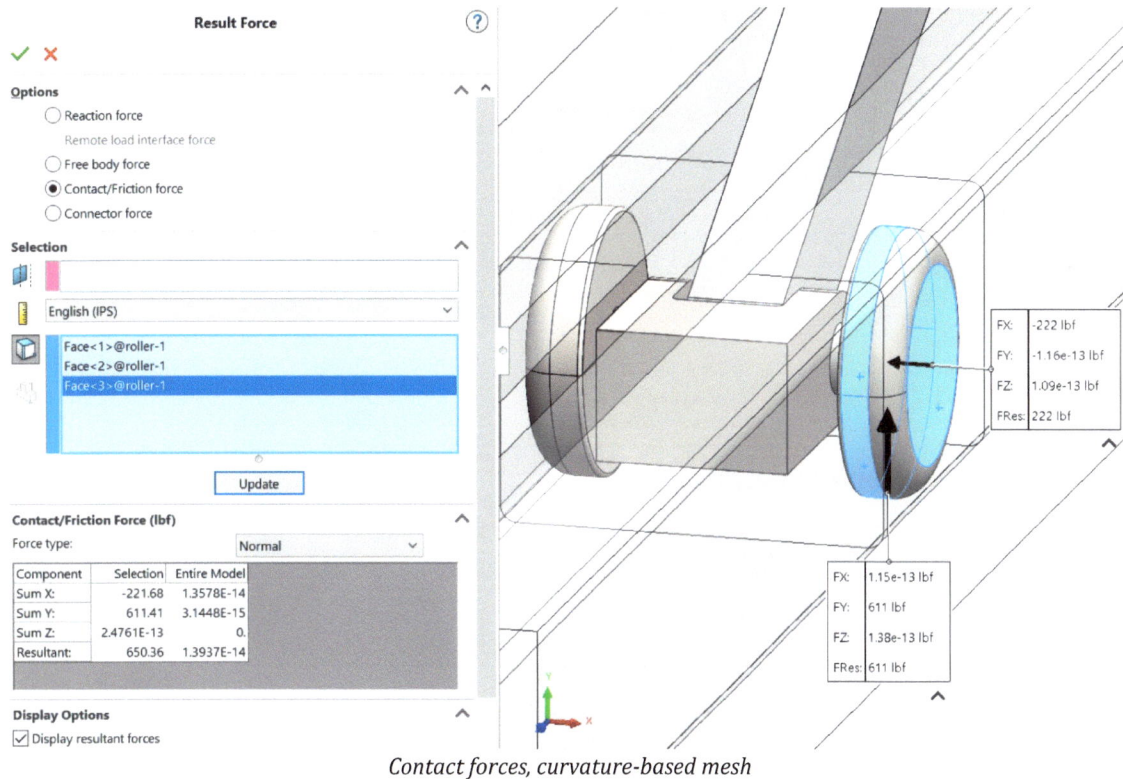

Contact forces, curvature-based mesh

Tabulation of the studies highlights a few points. Splitting the surfaces let us get a faster run from a more refined mesh. We let the solver worry only about select local faces in contact.

Arm and roller			elem	run time	side force
1	global contact	standard mesh	64380	100%	100%
2a	manual contact w splits	standard mesh	76037	57%	411%
2b	manual contact w splits	curvature based mesh	145364	218%	600%

Force reporting turned out to be sensitive to the mesh. This is for a force perpendicular to the loading and resulting only from distortion in the solid parts, a subtle thing. When forces are taken up by discrete hardware like thrust rollers or wear pads the results should be more explicit.

Study 5: A-Arms on Shafts

Overview

We will look quickly at one more contrived model before getting into things that look useful.

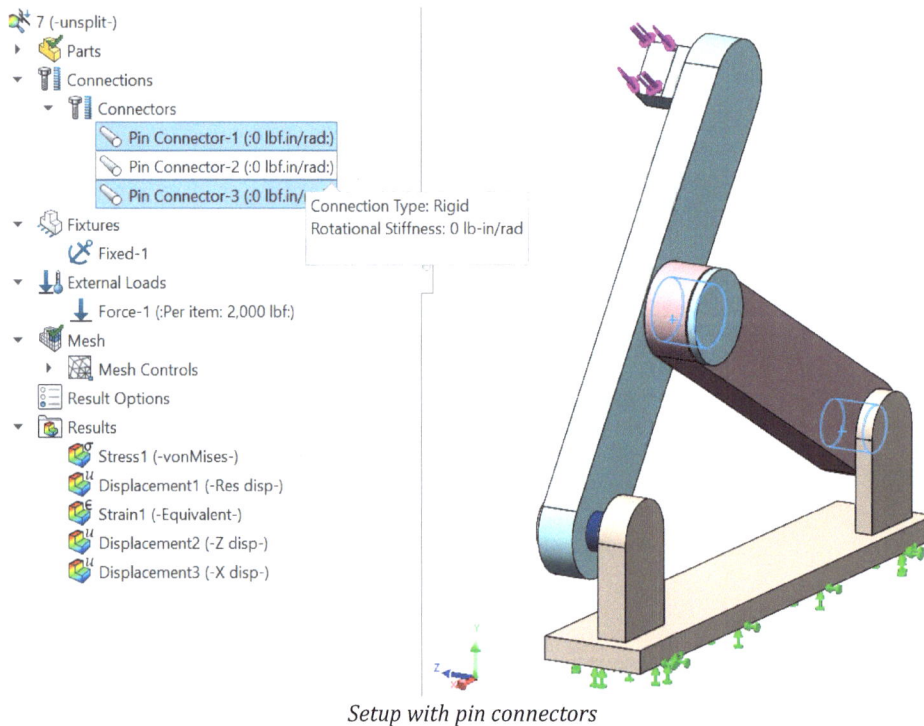

Setup with pin connectors

The arms are pinned together, solid pins integral to each arm model. The load is applied off to one side of the long arm, to impose some sideways bending through the arms. We start with rigid pin connectors.

Objective

Stress in the supporting towers and their shafts are desired. We want to see the effects of different sets of face splits and contact types.

Challenges

Simple models of the connections may not give good results. Better models of the connections may take a long time to setup and solve.

We will focus on the stress result through the end of the short arm and the thinner 'tombstone' support. The study runs in seconds, but the result is suspect. There is no stress through the shafts. The connectors are changed to the "distributed" pin.

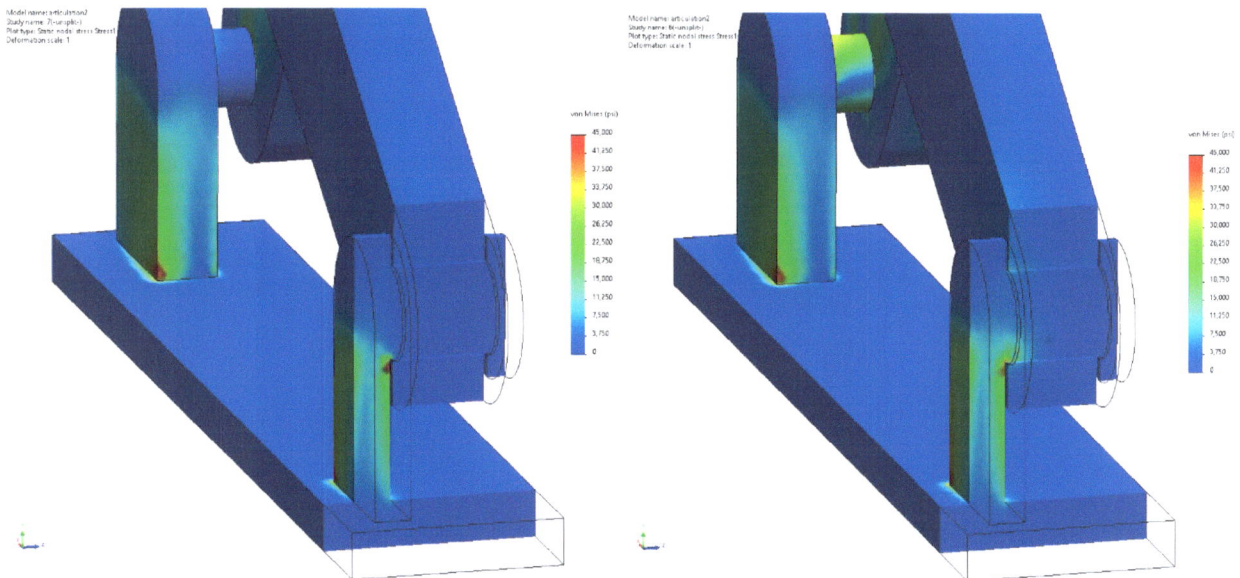

Result, rigid pins vs distributed pins

Both mounting pins show a bending stress field with the distributed pins. Bending stress in the support is more clear.

Again, we try global no-penetration contact. The result is livelier. The short arm can slide into lateral contact with the support. Stress details are hotter. Setting up the run with manual contact sets gives a slightly faster run time.

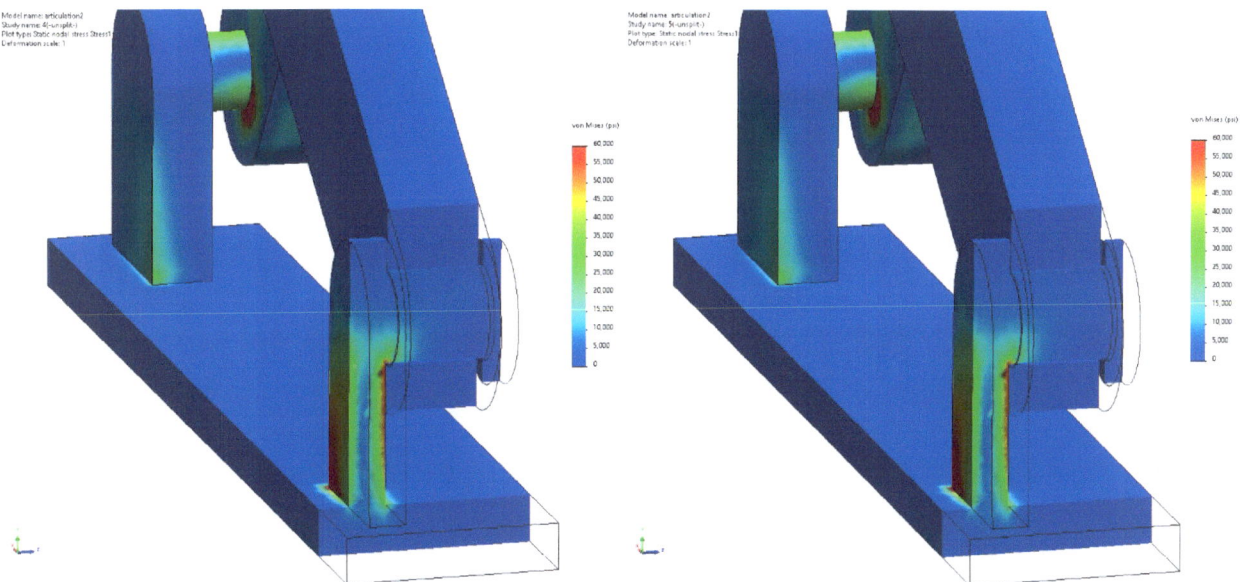

Result – global no-penetration contact vs manual contact

We are going to split the surfaces to support manual contact setup.

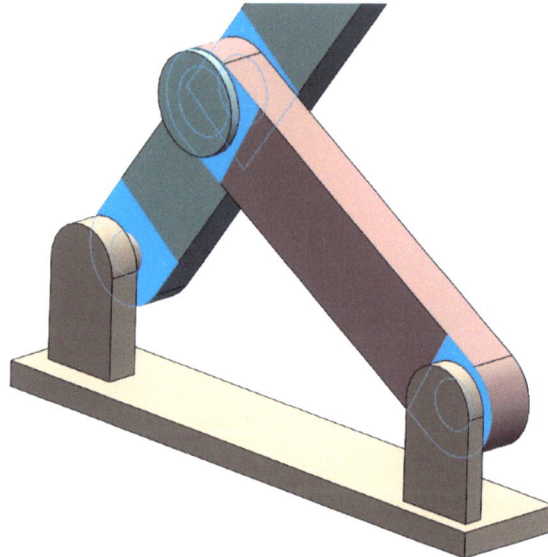

Model with split surfaces highlighted

But what happens if we run the split model in global no-penetration contact?

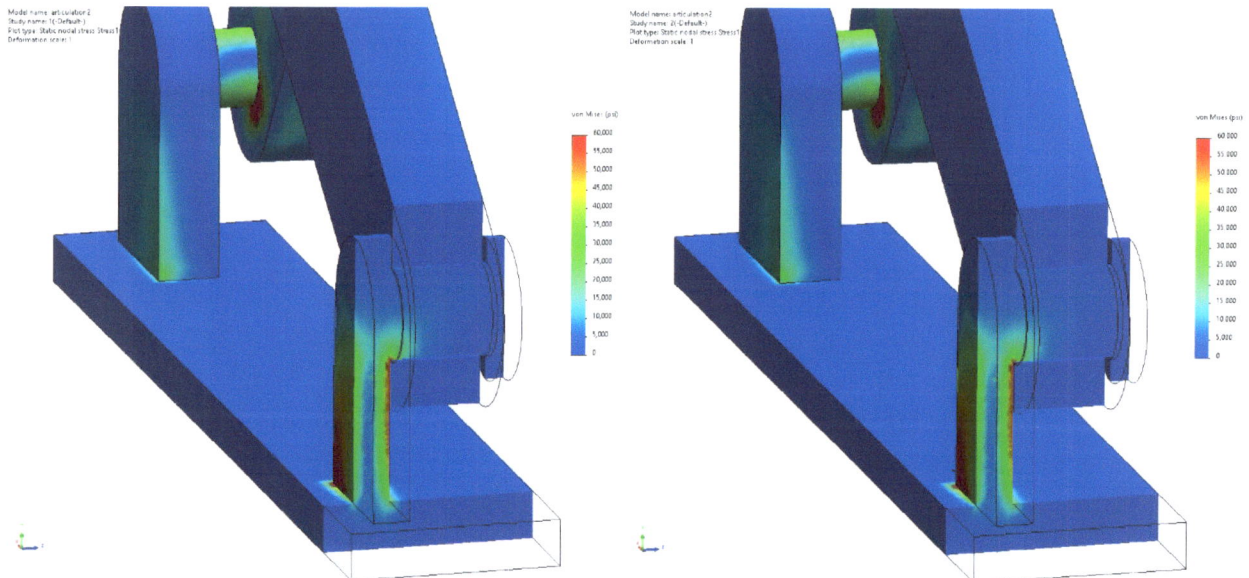

Result – split model in global contact vs manual contact

We get the same result, but the mesher was more efficient with the split model, made a smaller mesh, and the result was a faster run. The solver may also have an easier time sorting out global contact with the pre-split surfaces.

Up to now we have not gone into any detail about no-penetration contact sets. There are important options for contact sets. [A simulation option must be set to "Show advanced options for contact set definitions".]

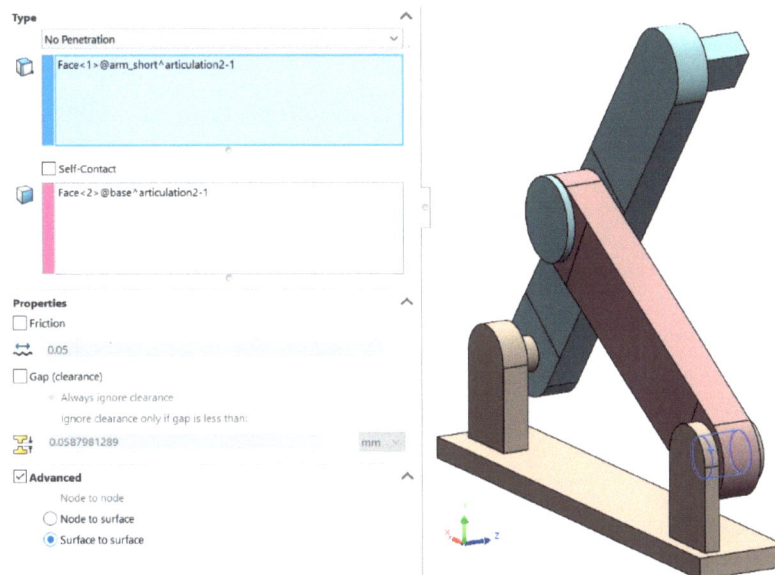

Contact set definition

The options usually available are "Node to surface" and "Surface to surface". Solidworks documentation explains only that surface-to-surface is more accurate but takes longer to solve. Long experience, mostly with round rollers and flat pads in contact, has led us *always* to use surface-to-surface contact.

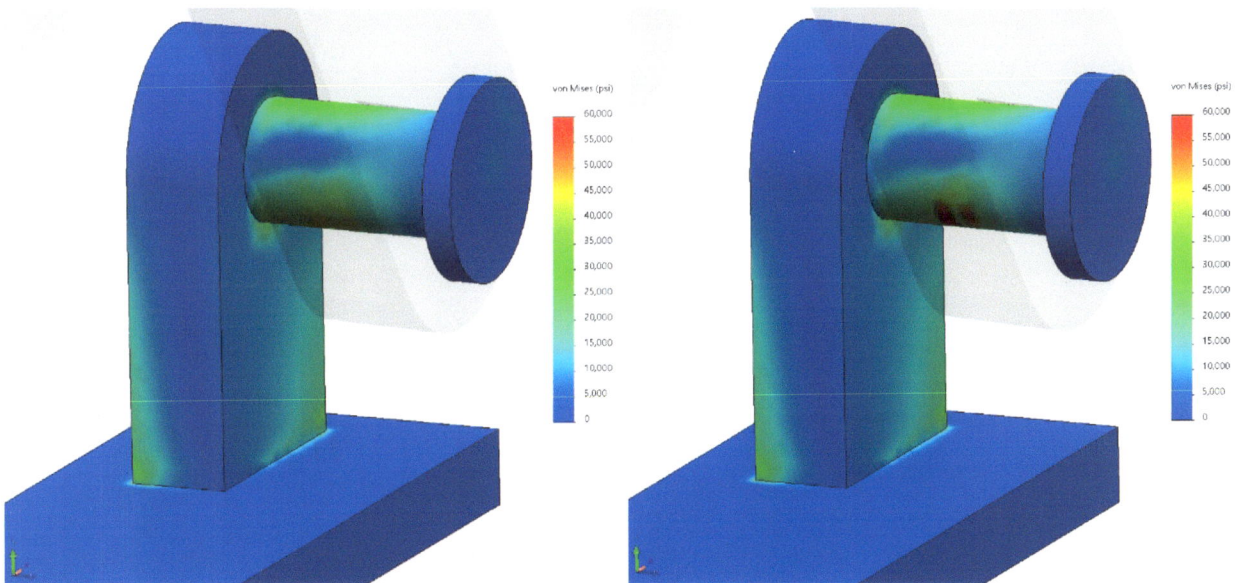

Result – Surface-to-surface vs node-to-surface contact

These plots show the rear support shaft. The stress result differs in small surface details. They are probably not important here. In other studies, we find a larger difference or a better chance at getting a stable solve with the surface-to-surface contacts. The cost here was less than two percent of total run time.

Shafts in Bending (articulation2)	model		elem	run time
pins - rigid	unsplit	7	82966	2%
pins - distributed	unsplit	6	82966	3%
Global contact, unsplit surfaces	unsplit	4	82966	100%
manual contact - surf-to-surf, unsplit surfaces	unsplit	5	82966	99%
Global contact, split surfaces	split	1	76293	89%
manual contact - surf-to-surf	split	2	76293	86%
manual contact - node-to-surf	split	3	76293	84%

Summary of studies

In this example we find running full contact gives the most complete picture of mechanical motion and resultant material stress. Small differences in run time are seen with different contact setups on the same size mesh.

Study 6: Hydraulic Lift

Overview

A compact lift is designed with a single hydraulic cylinder. The moving parts are pinned to a machined casting which in turn can be welded or fastened to something else, like the back of a truck. The mechanism is similar to that which lifts and lowers a snow plow.

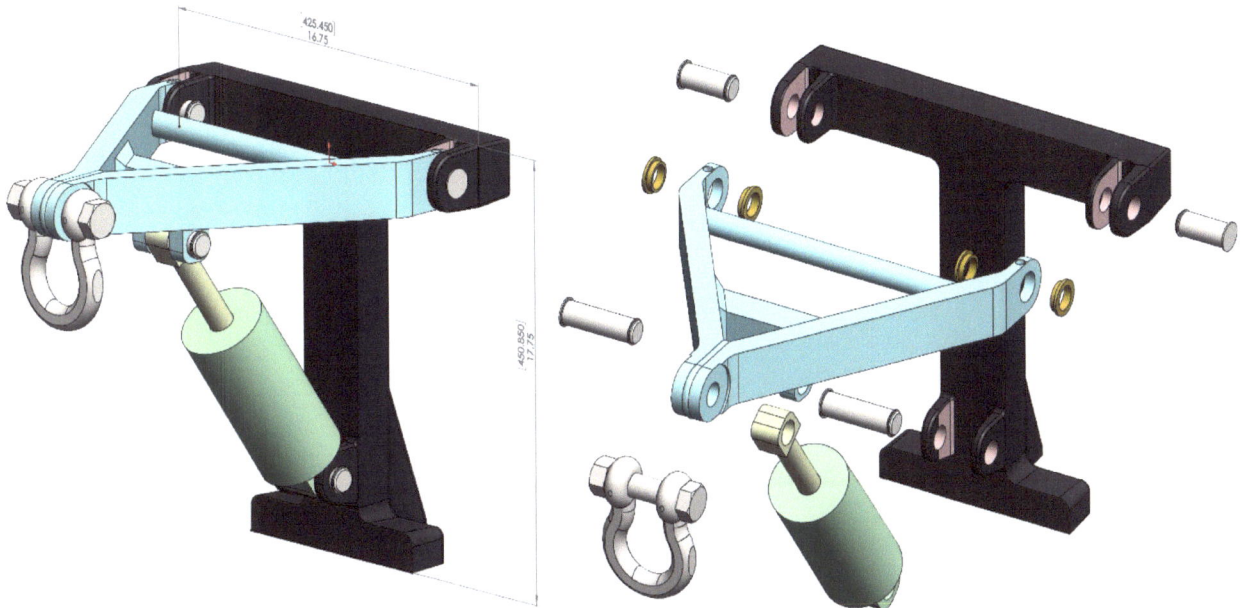

Lift assembly, outline dimensions and exploded view

Objective

We want to know inter-component forces and find high stress areas with the lift at different positions. We also want to check the structure at maximum hydraulic pressure.

Challenges

The hydraulic cylinder can both resist and apply force. We must find a way to represent it that works in different modes and positions.

Model and Initial Setup

The model is imported and found to have close clearance between all components, which are mated in multiple position configurations.

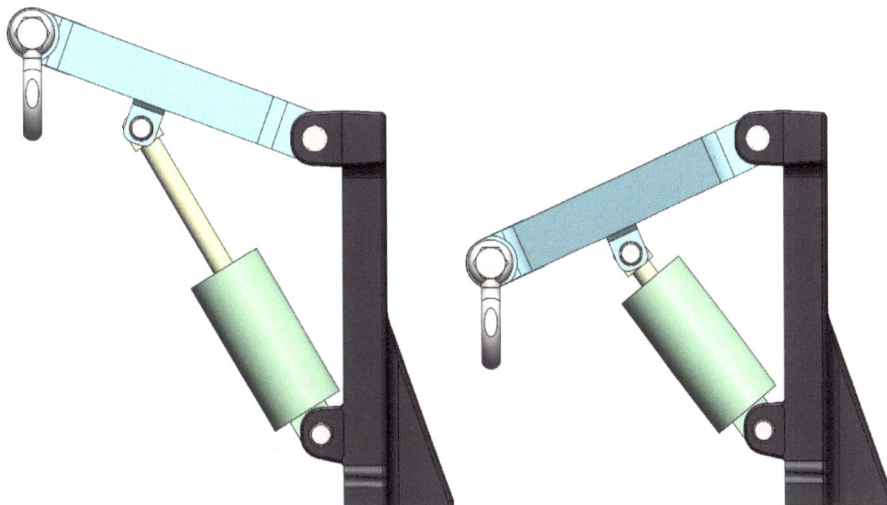

Model at limits of travel

Most connections are by plain headed pin and snap ring (the snap rings are modeled as solid parts of the pins). A section through one of the two rear arm joints shows that this connection has two bushings and provision in the arm for a threaded grease fitting to be installed.

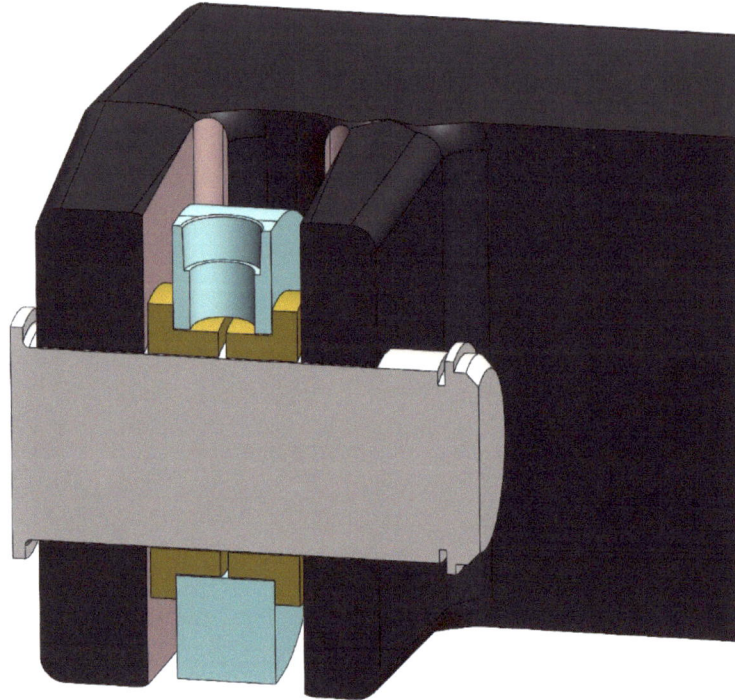

Section through arm joint

Hydraulic Cylinder

A single-stage hydraulic or pneumatic cylinder will always have at least a shell and piston. On each end we could find a simple rod, a single clevis, a double clevis, a ball joint, a threaded end, or even a bolt flange. For this assembly, a simple model with a single clevis on each end has been set up.

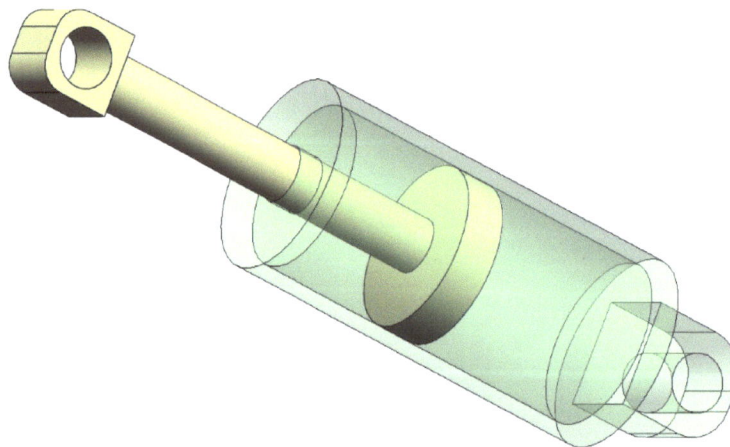

Cylinder model

The cylinder model has very little detail. The cylinder is considered a purchased part of this assembly and we are not trying to re-do the work of its designers. For our purposes all we need is for the model to be two separate parts (this could be a flexible subassembly) which have appropriate end shapes and coaxial features in the middle to carry the intended motion. It is helpful in design if the limiting piston and shell dimensions are accurate, so that limit mates can be used to show how the machine moves, but this is secondary in the simulation work.

What we need in simulation is effective end connections and a definition of the piston being located in the shell. The most literal way to do this is to put contact conditions everywhere. Here one clevis end is put in no-penetration contact with the lower pin and both inside faces of the base clevis.

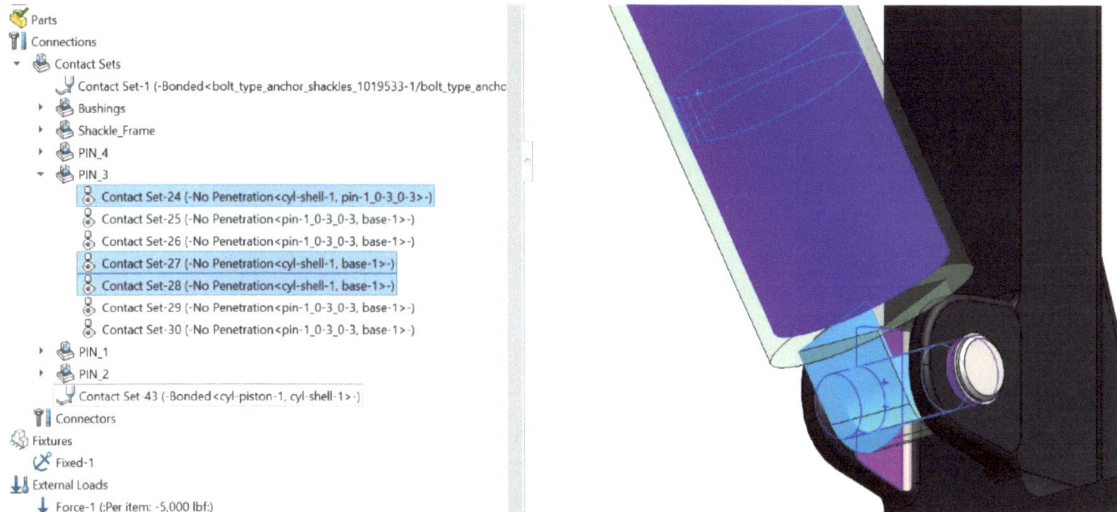

Contact definition

Then the piston outer diameter is put into *bonded* contact with the inside of the cylinder shell. This is appropriate if the cylinder is the holding element of the lift.

A similar effect can be had using a pin connection with 'retaining ring'. The piston outer diameter here is pinned to the cylinder shell.

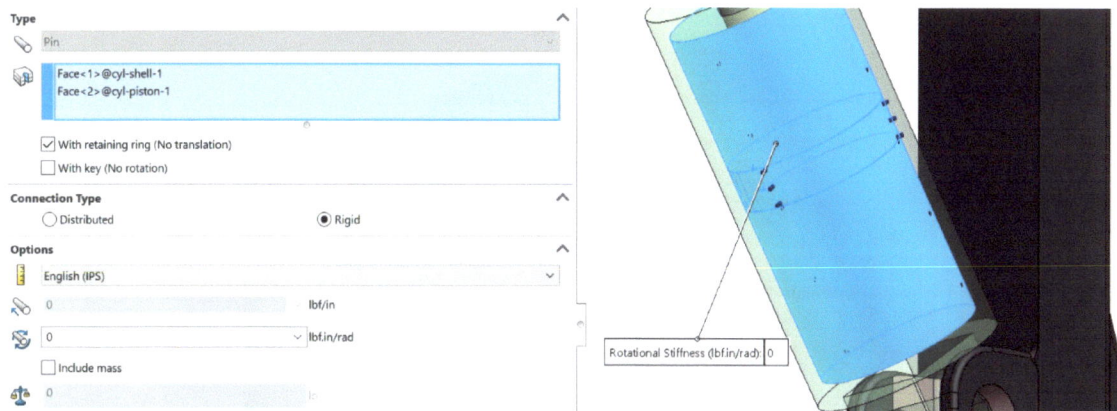

Pin definition

Note the options in the pin connector for input of linear or rotational stiffness. A lot of interesting things can be done here, but for now we assume the cylinder is closed off and the hydraulic fluid is incompressible. The piston cannot slide inside the shell. We leave rotation free, as most cylinders do allow rotation and we want the ends to move freely into realistic positions.

Initial Run

The customer has specified a load rating of 5000 pounds and explains that it can swing around a little. A reference plane is defined at12 degrees from horizontal and a force applied to the shackle at that angle. The assembly will see over 1000 pounds of lateral force and the cylinder may see some side loading.

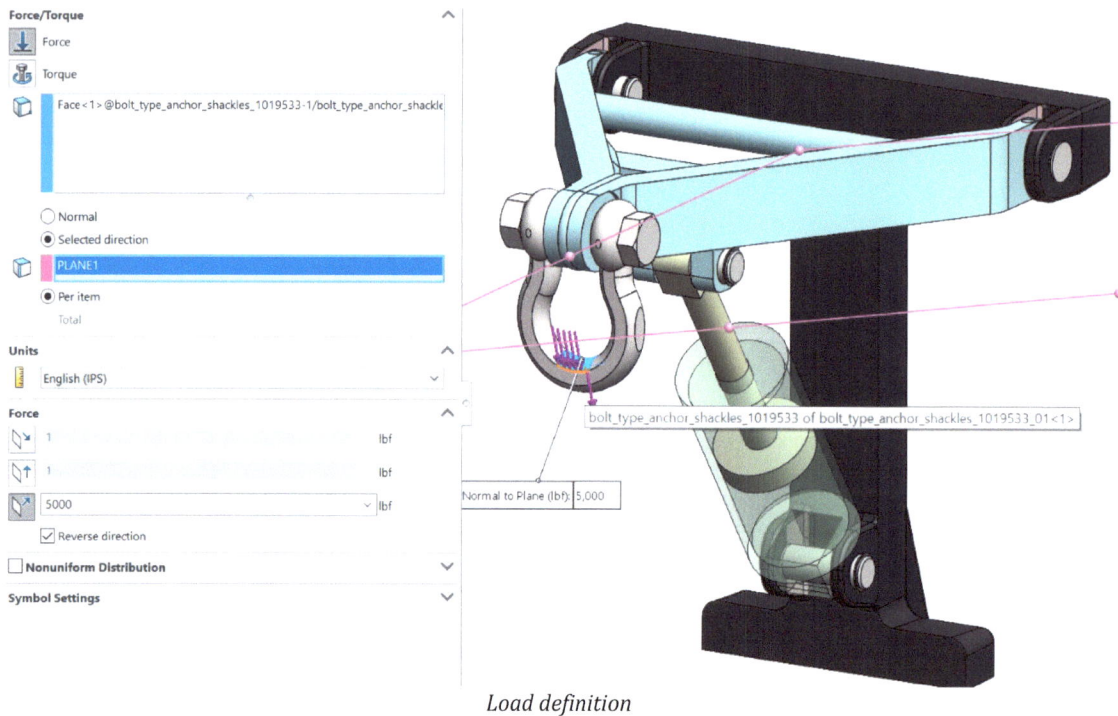

Load definition

Since the model looked pretty good, we define a study with a fast coarse mesh and a global no-penetration contact condition, hoping to get a first result quickly. The cylinder piston is bonded to the shell. It takes near zero effort to try this as a first pass. That study runs while we set up a "real" study, with defined contact and/or connectors.

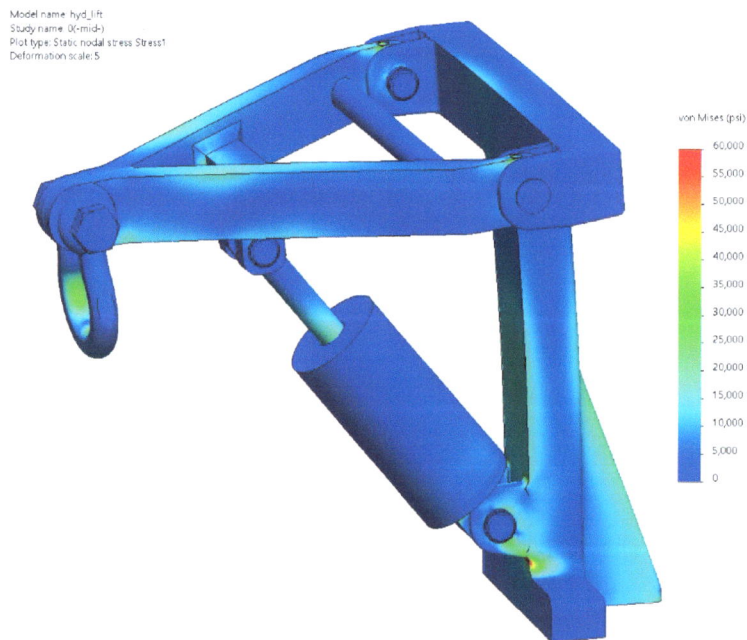

Result of quick study [5X deflection]

The initial result looks reasonable and the deflection is all in the expected directions. It took almost no time to set up and fifteen minutes to run.

Full Contact Setup

In those fifteen minutes work was started on a full contact setup. Every pair of faces that is expected to touch is identified and a separate contact set defined. Here four contacts are highlighted for one of the bushings – inside and outside diameters and both sides of the flange.

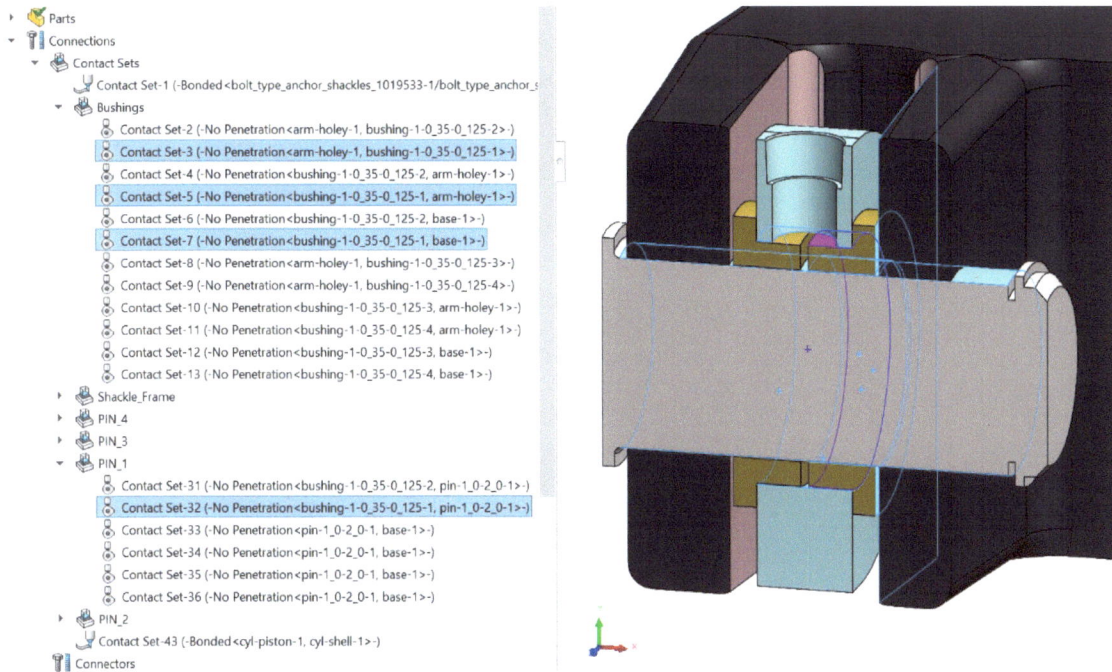

Contact sets for bushing 1

Spelling out the contacts in discrete pairs will pay off later when we ask what the contact forces came out to be. If we define the contacts, only those surfaces will be involved and there is no guessing what the solver did. It is also a little faster to solve and sometimes much more memory efficient if only select surfaces are picked out for contact.

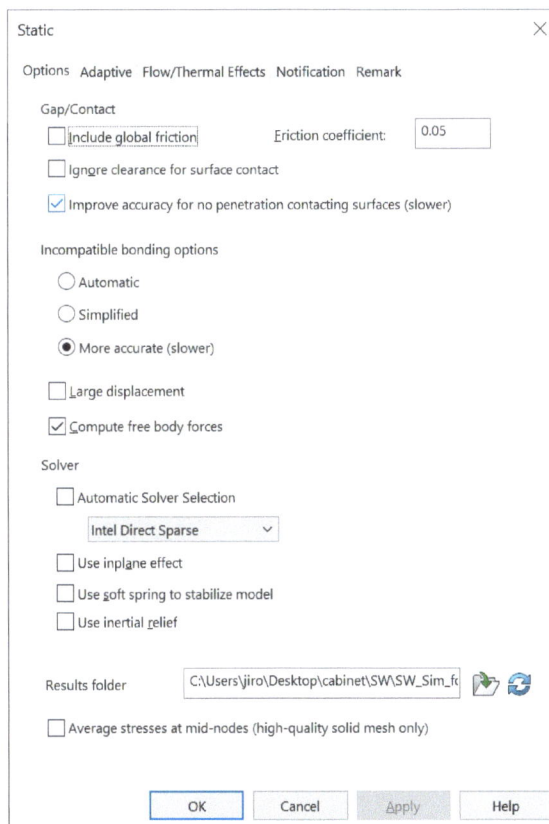

Study properties

One must remember before starting to open study properties and check "Improve accuracy for no penetration contacting surfaces". This pre-selects surface-to-surface contact for each set, which works much better than the other options. You can always go back and edit the advanced properties of each contact, but that is an annoying chore when it could have been checked off in advance.

It is also highly recommended to group contacts into folders.

A few mesh refinements are defined at the welds and around the grease fitting holes.

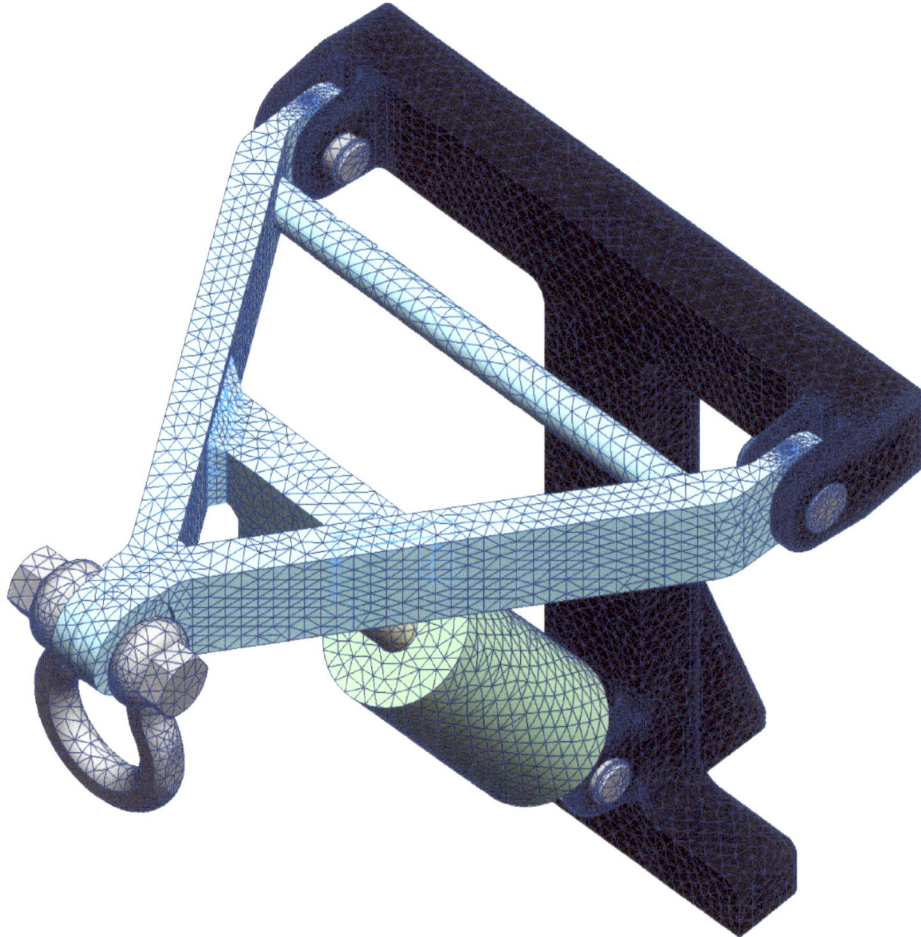

Mesh on detailed study

The setup is otherwise unchanged. The overall result is similar.

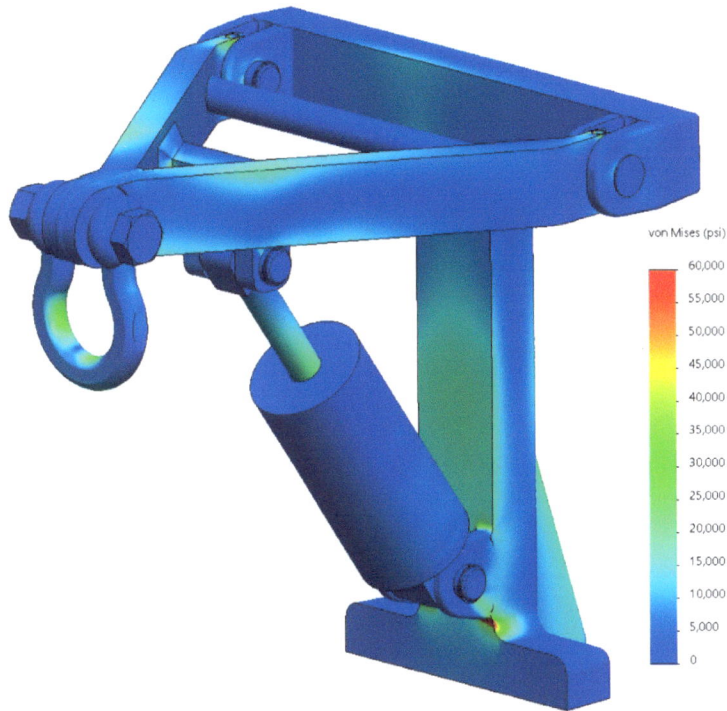

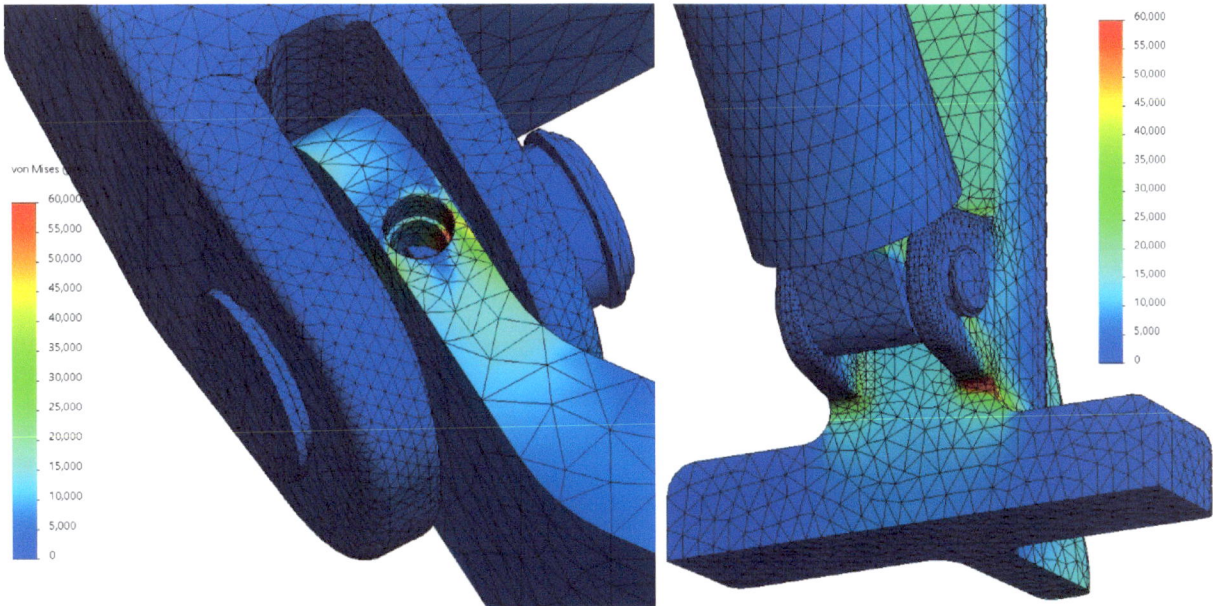

Result from full contact study

A couple hot spots show up clearly. At the right (user viewed) arm grease fitting hole the thin wall and local curvature combine to make a stress concentration. [The original purpose of this model was to demonstrate this effect after a real failure was seen in a different sort of mechanism.] Also hot is one ear of the cylinder support on the base casting.

The results are asymmetric. Since the load had a lateral component this is expected. A plot of lateral (X) displacement shows how things moved. The shackle had a lot of free play and it did slide fully to the side until making contact. The cylinder looks to have moved a bit in the lower clevis. The A-arm moved less, as it was close fit in the bushings, but it deformed a good bit at the tip.

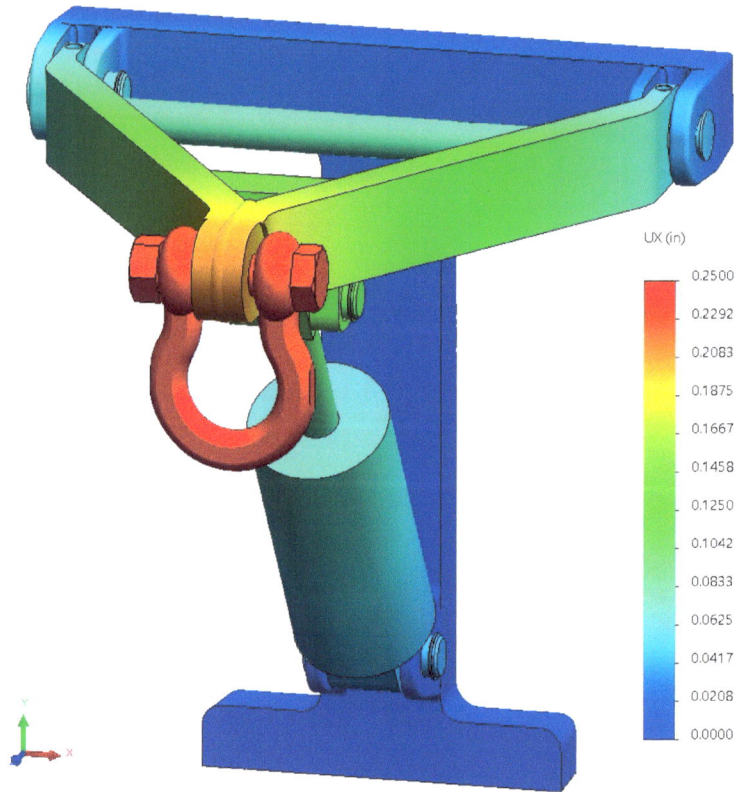

Lateral displacement

We wonder how the load is being transferred through the parts. Here is the payoff of defining the contacts in discrete pairs. The "List Result Force" function is used to access force predictions on select faces.

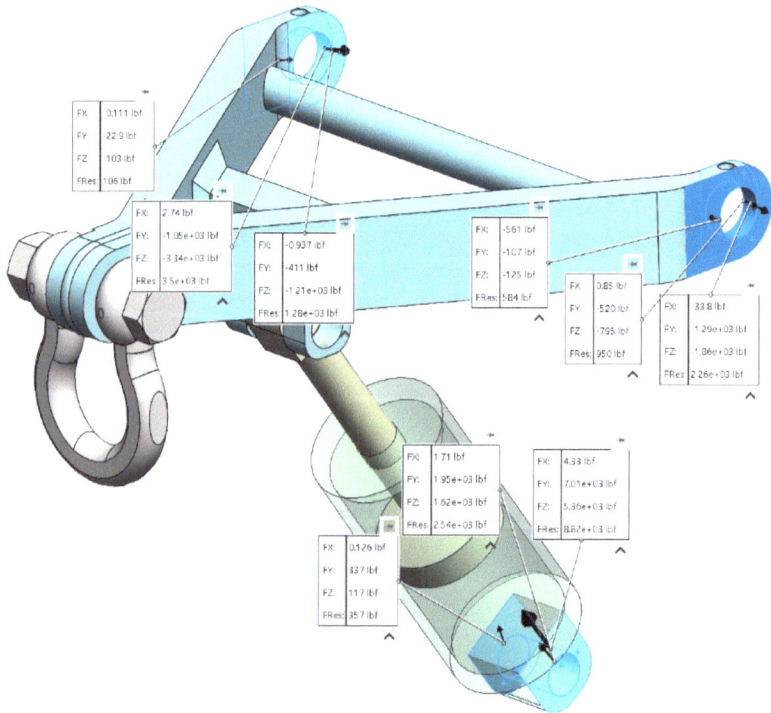

Contact forces

Here the rear frame and pins are hidden and all the faces which were in contact with them are selected. We expect this subset of parts to be in static balance, so we add up the lateral forces to check. The X-direction forces add up to 1070 pounds. The applied force should put in 1040 pounds. We consider this good agreement.

Vector sums are shown and we can pull out the Y and Z portions to calculate shear forces on the pins, which are the forces trying to rip open the ends of our arms. The right arm sees over 3500 pounds force, the left just 2246. Compressive force in the cylinder is 8824 pounds in this position.

Setup with Pin Connectors

What if the design was in conceptual stages and all we really wanted to know was the cylinder force in different positions? We know that running with pin connectors can give results really fast. So, the study is duplicated, and all those hard-earned contact definitions are deleted.

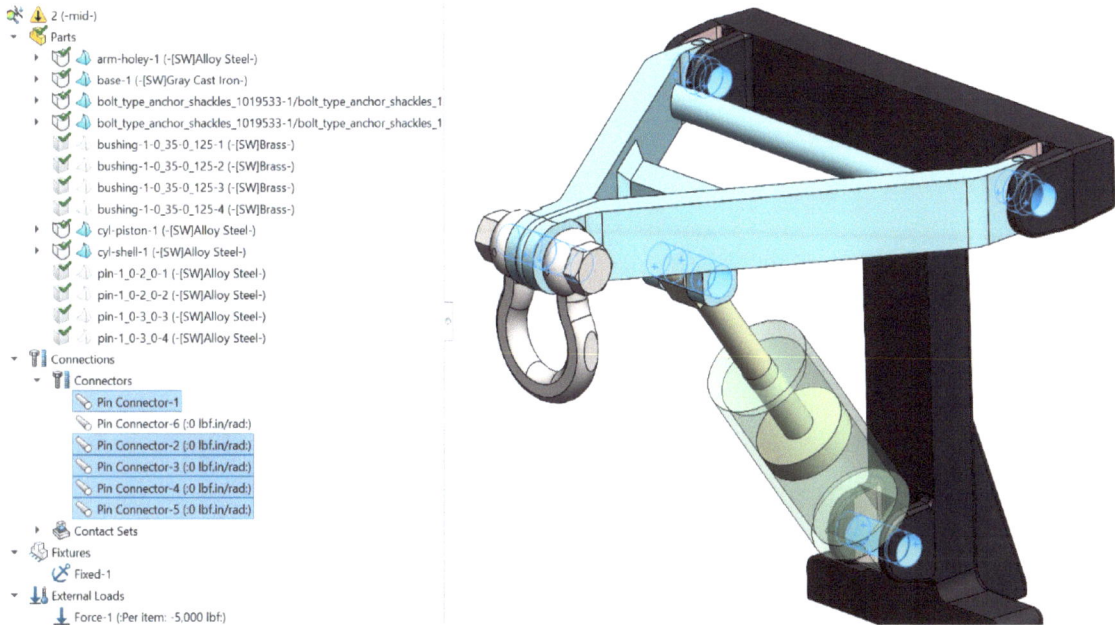

Setup with rigid pin connectors

All the pins and bushings are "excluded" from the study. Simple rigid pin connectors are defined at each joint. The cylinder piston is also pinned to the shell. Run time with the same mesh settings is 70 times faster than with the full contact setup.

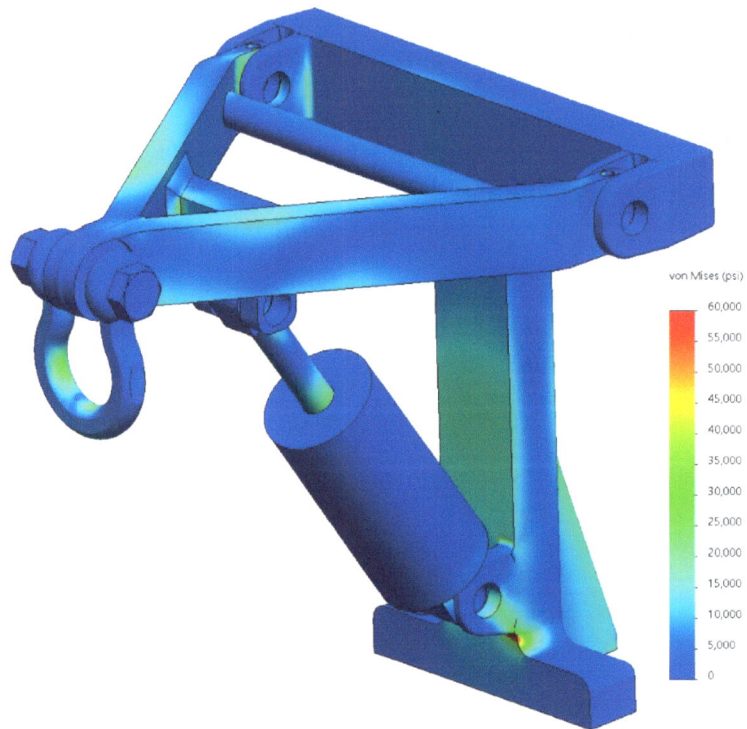

Result with rigid pin connectors

The overall result looks good. The lower hot spot still shows up plainly. But something is missing.

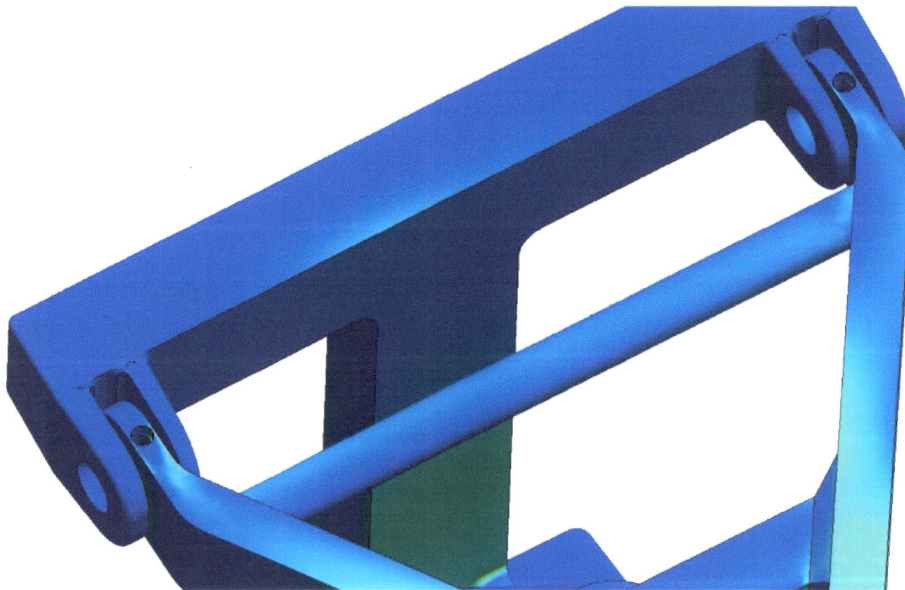

Neither side of the A-arm shows a hot spot at the thin walls. As we learned earlier this is a consequence of using the rigid pin connector, which locks surfaces on pivoting but otherwise unmoving cylinders. No radial stress can develop at those surfaces.

It's easy enough to change the pin connectors to "distributed" so we try that.

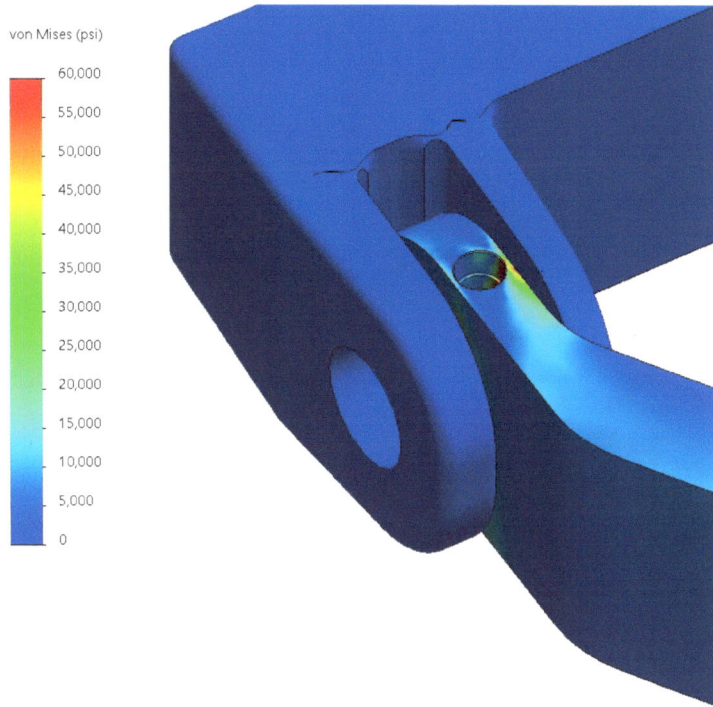

Result with distributed pin connectors

Now the stress hot spot at the grease fitting hole is again revealed. The cost was a near doubling in run time – from 23 seconds to 43 seconds. At that cost we can stick with the distributed pins. Stress plots can be compared to the full contact run and only fine details around the connections are any different.

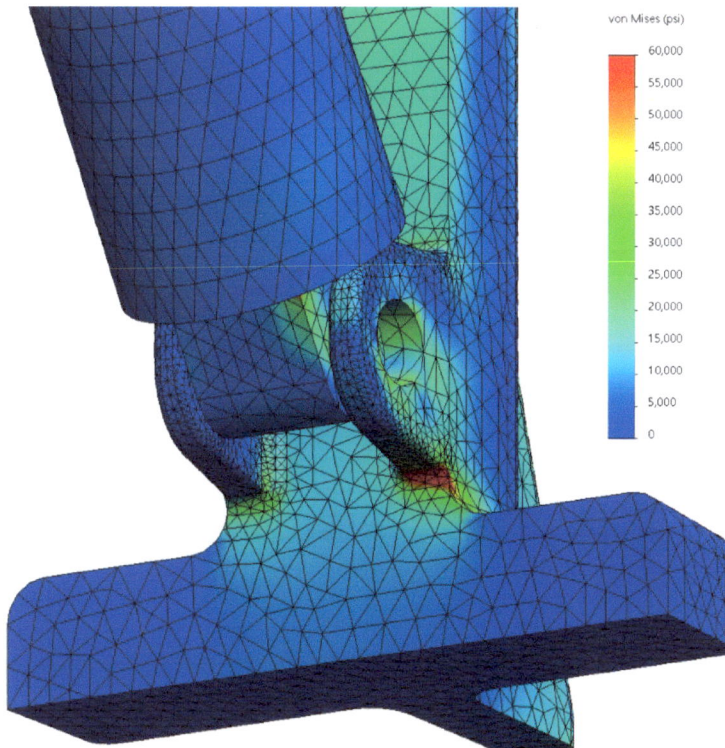

Result detail with distributed pins

The lift can be loaded at different positions. With a run time under a minute, we can easily afford to run a full study in each position and look at how the forces change.

Cylinder Force

Result forces and plot of cylinder force

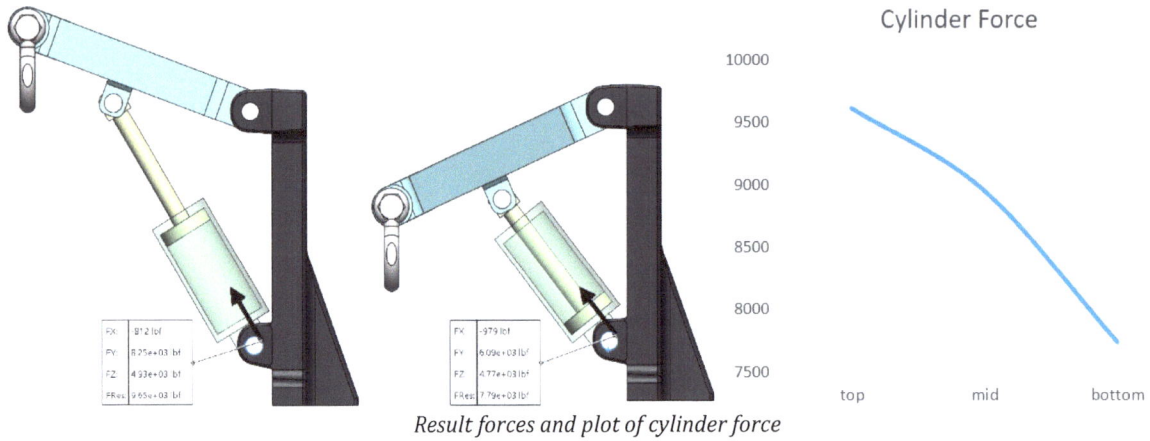

Alternate Cylinder Setups

We looked at the cylinder parts being in bonded contact or joined by a pin. Here is another way to set it up, using a spring connector. The cylinder piston is still 'pinned' to the shell, but the retaining ring condition is tuned off, so the piston can slide in the shell freely.

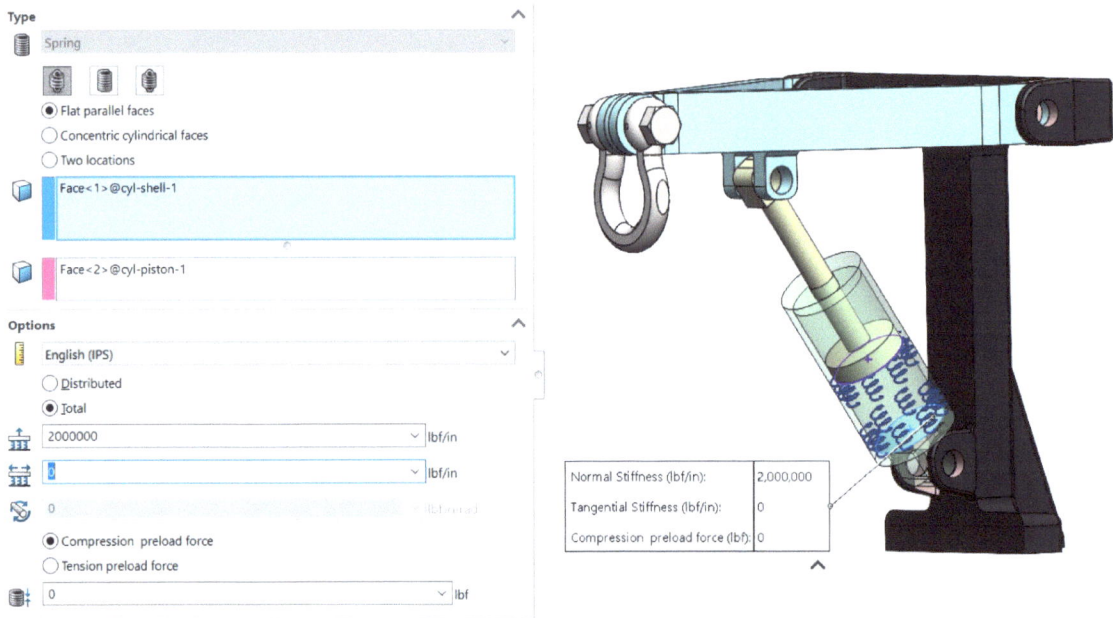

Spring setup inside cylinder

A spring is defined between the piston and inside bottom of the shell. The axial stiffness in the spring is an estimate of the compressibility of the hydraulic fluid. As mentioned before, we also could have put some compressibility in the pin connector. We also could probably eliminate the pin connector and use a very high lateral stiffness in the spring. But this setup is plenty quick and effective.

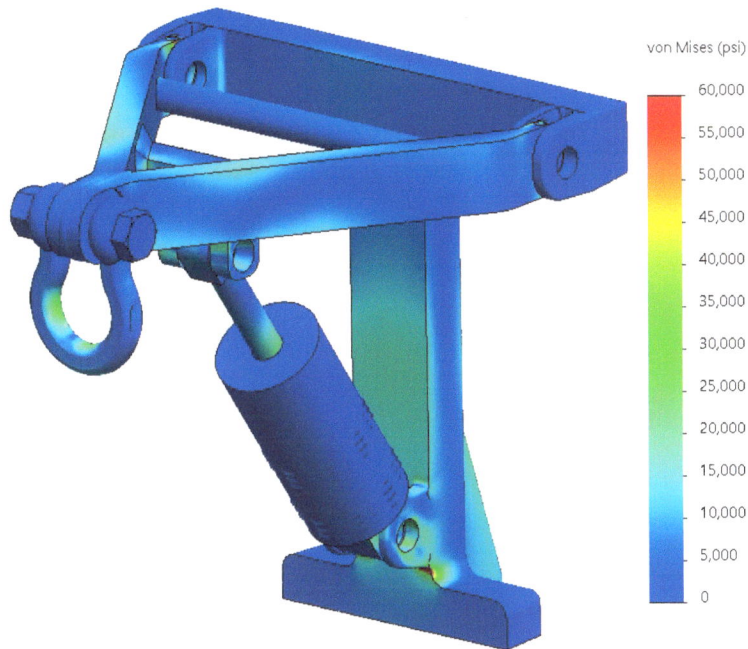

Result with compressible spring

What if we were not concerned at all about the cylinder at this stage of the design? All we really need to do is link the two end pin areas with something that will transfer force in the appropriate line. That something can be the "link" connector.

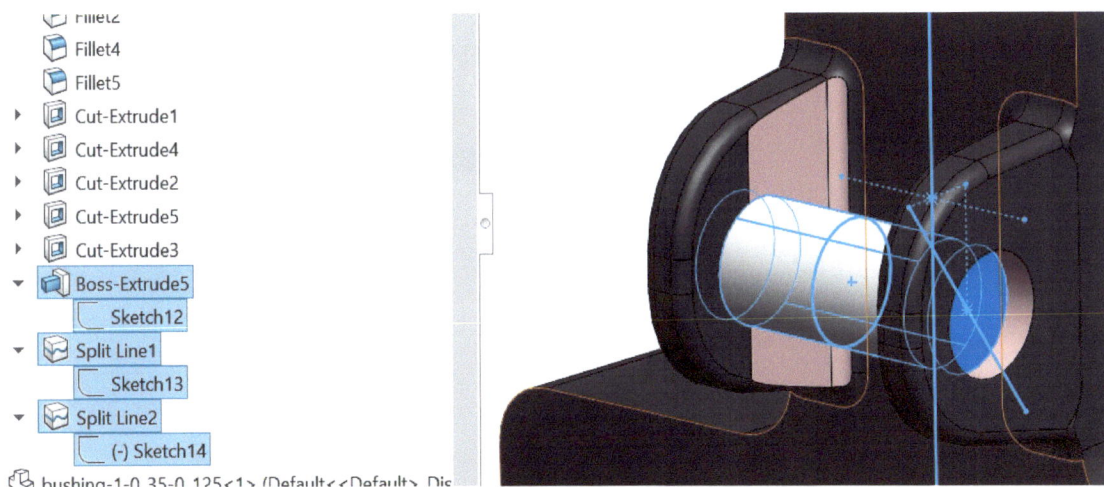

Model edits to make split pin

The base and the A-arm are edited so that a solid pin becomes part of each body. The surface of each new pin is split so that there is a vertex in the center of where the cylinder would have pushed. A link connector is then defined between the two new center points.

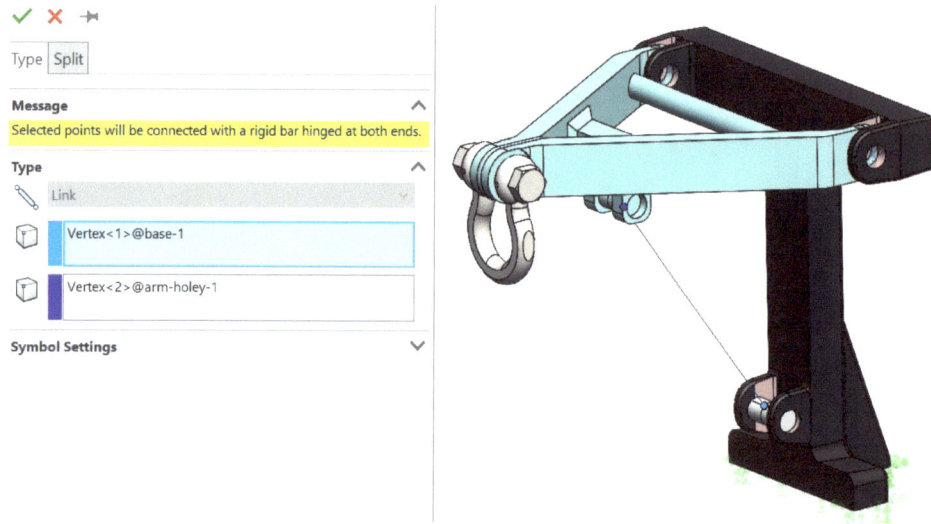

Definition of link connector

The definition of a link connector is two vertices. There are no other options.

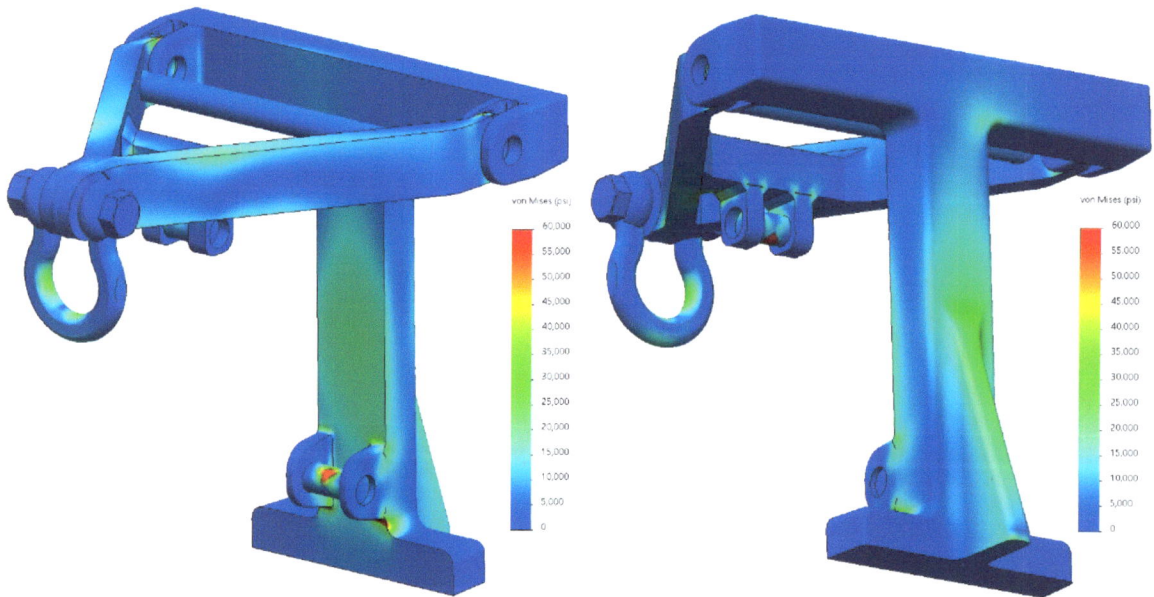

Result with link connector

The overall result is still close to those with more real detail. Of course, stresses around the cylinder mounting pins are different, as the pins are now parts of the solids and the link connector makes a large stress concentration at its point contacts.

One other important thing about the link connector is evident here. The link cannot transmit any lateral force.

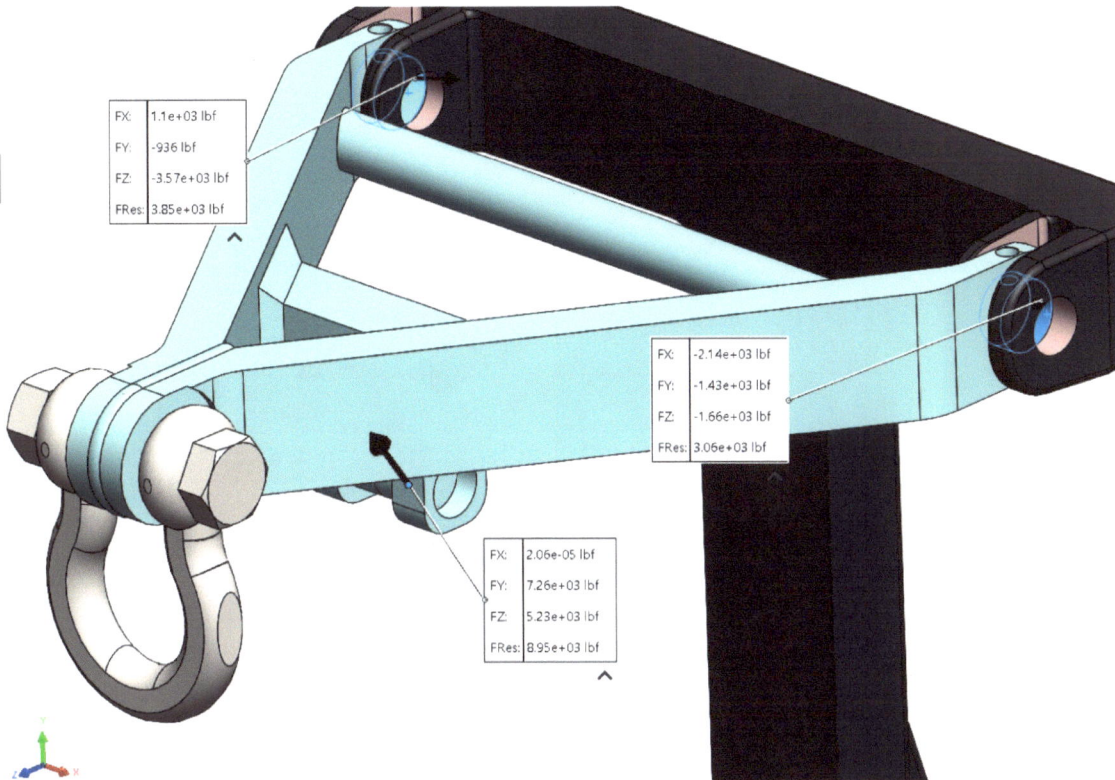

Resultant forces on arm with link connector

The entire expected 1040 pounds of lateral force must be made up between the upper joints. This will lead to an over-prediction of net force on one side and a low prediction on the other.

Cylinder as Driving Element

In real world use there is always a chance this mechanism can be caught on something that just isn't going to move, or the user could unknowingly overload it. We must check the structure against the maximum force the cylinder can generate. We could increase the applied load and iterate to get the expected max cylinder force, but here we will use the features of the cylinder to make a prediction directly.

First the setup is reversed so that the original applied load becomes a fixture. The center of the shackle is restrained in the horizontal plane.

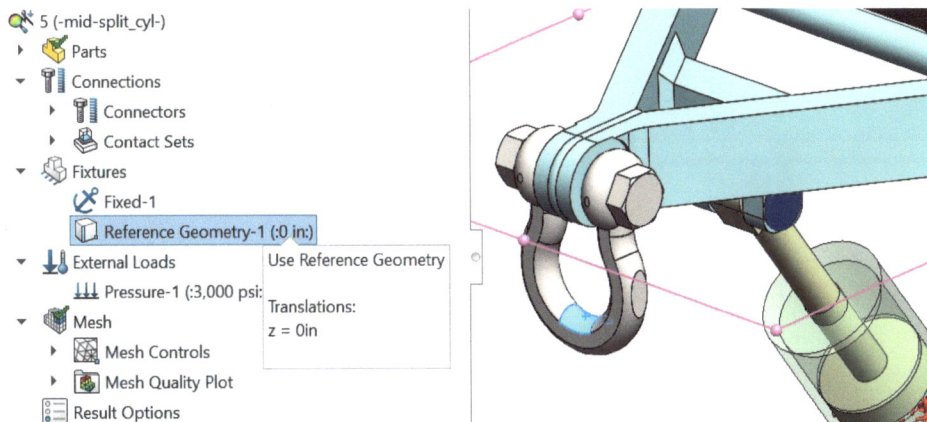

Setup with new fixture

Then we must pressurize the cylinder. This is done by literally applying pressure inside the cylinder. First the inside face of the shell is split so that the portions around, above, and below the piston are separate. The piston is put into no penetration contact with the shell, just in the small adjacent band.

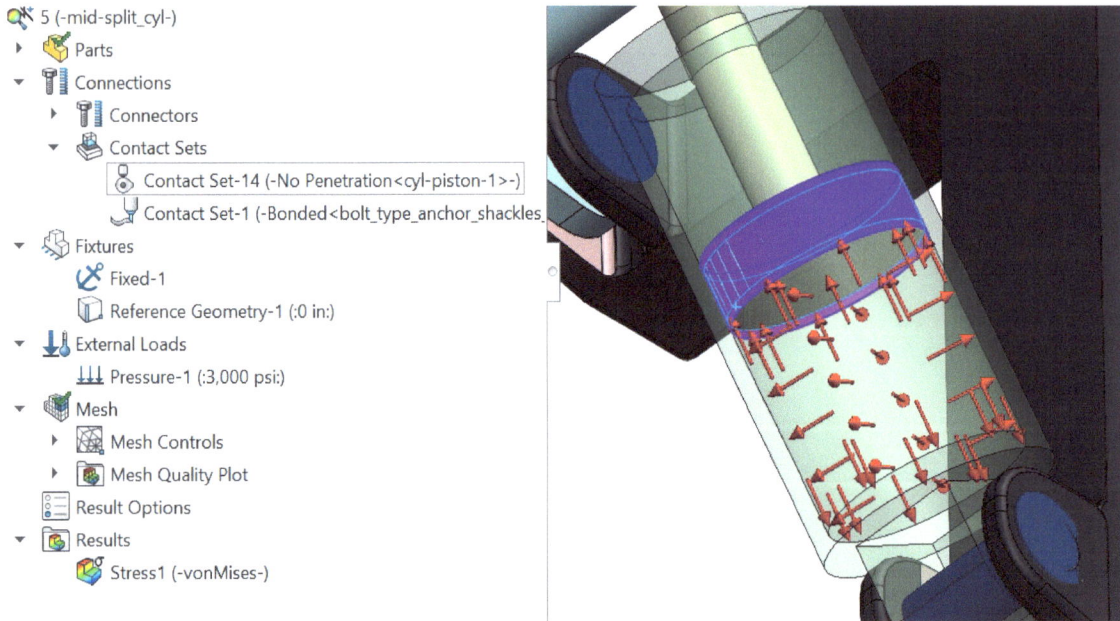

Setup of pressurized cylinder

The hydraulic system is planned to have a blow off at 3000 psi, so that pressure is applied to the inside faces of the cylinder on the high pressure side. The result is dramatic.

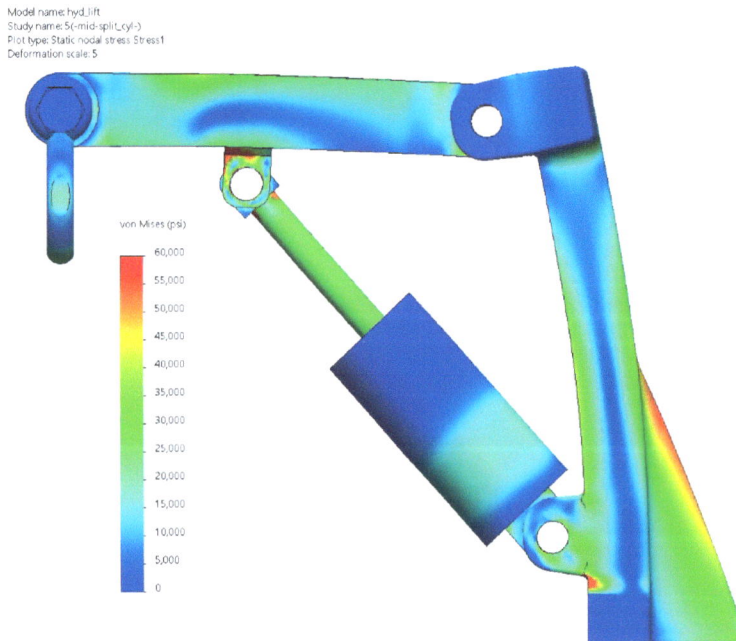

Result at max cylinder pressure (5X displacement)

The resulting cylinder force of over 21,000 pounds matches expectations. This force bends the structure noticeably and over stresses the previously noted hot spots and most of the welds. We can recommend a lower pressure or a smaller cylinder.

Tabulation of the runs here includes force data. Most runs correlate well to each other in force balance. Run time for all the models with synthetic connectors is much lower than the full contact study, but the force and stress details on the small hardware like the bushings cannot be got from the faster runs.

	Hydraulic Lift	Arm R	Arm L	Cyl		ARM REAR RIGHT			ARM REAR LEFT			Cylinder bottom				
		FX	FX	FX	sum	FY	FZ	vec	FY	FZ	vec	FY	FZ	vec	elem	run time
1	full manual contact	1.692	-593.95	-478	-1070	-1050	-3340	3501	-1290	-1860	2264	7010	5360	8824	183071	00:27:10
2	rigid pin connectors	1680	-1620	-1100	-1040	-1310	-3250	3504	-1060	-1980	2246	7260	5230	8948	148330	00:00:23
3	distributed pin connectors	1560	-1680	-921	-1041	-1230	-3260	3484	-1130	-1970	2271	7260	5230	8948	148330	00:00:43
3-top		1470	-1700	-812	-1042	-1880	-3050	3583	-1480	-1880	2393	8250	4930	9611	149661	00:00:45
3-bot		1400	-1460	-979	-1039	-454	-3040	3074	-741	-1730	1882	6090	4770	7736	149113	00:00:43
4	cyl fluid as spring	1560	-1680	-924	-1044	-1240	-3260	3488	-1130	-1970	2271	7260	5230	8948	148330	00:00:43
5	cyl split and pressurized - pushing				0			0			0	17200	12400	**21204**	147806	00:02:38
6	cyl is link, pin conns	1100	-2140	0	-1040	-936	-3570	3691	-1430	-1660	2191	7260	5230	8948	141992	00:00:28

Study 7: Reach Handler on Carousel

Overview

This is where everything comes together. This assembly has rollers, ball joints, bushings, pins, tapered bearings, weldments, and castings. And it moves.

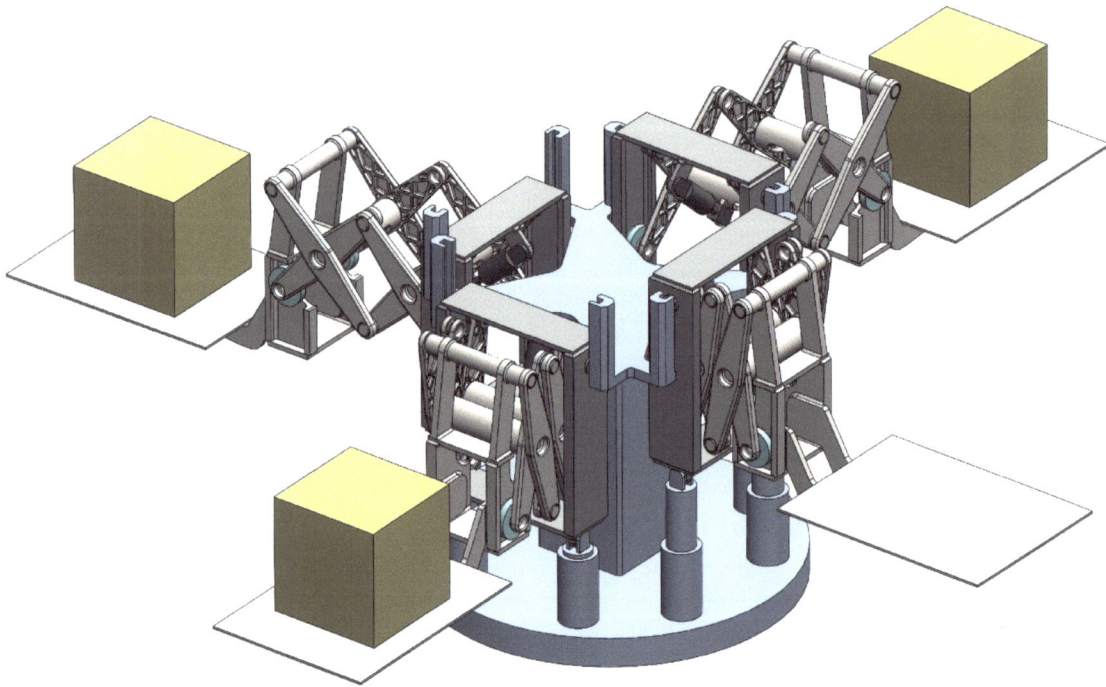

Four reach handlers on carousel

Four copies of the mechanism we want to study are mounted on a carousel. The carousel turns as each reach handler receives a load, extends, has the load removed, and retracts. We are not concerned here with the carousel, it is simply represented here to illustrate the lifting and turning arrangement.

Objective

Stresses in the cast arm design are a key consideration, along with stress predictions in the other parts. Some lateral force control hardware has been omitted from the design, so side forces on rollers and bushings are desired.

Challenges

We will need a way to input loading from the bulk motion of the whole machine. New kinds of hardware must be considered. The detailed casting geometry may present meshing and stress prediction problems.

The Assembly

The design concept for the reach assembly has arms that start with eight identical castings (A). The center and ends of each castings have different stubshafts welded to them or are machined out entirely. The center arms are paired together, welded to large tubes.

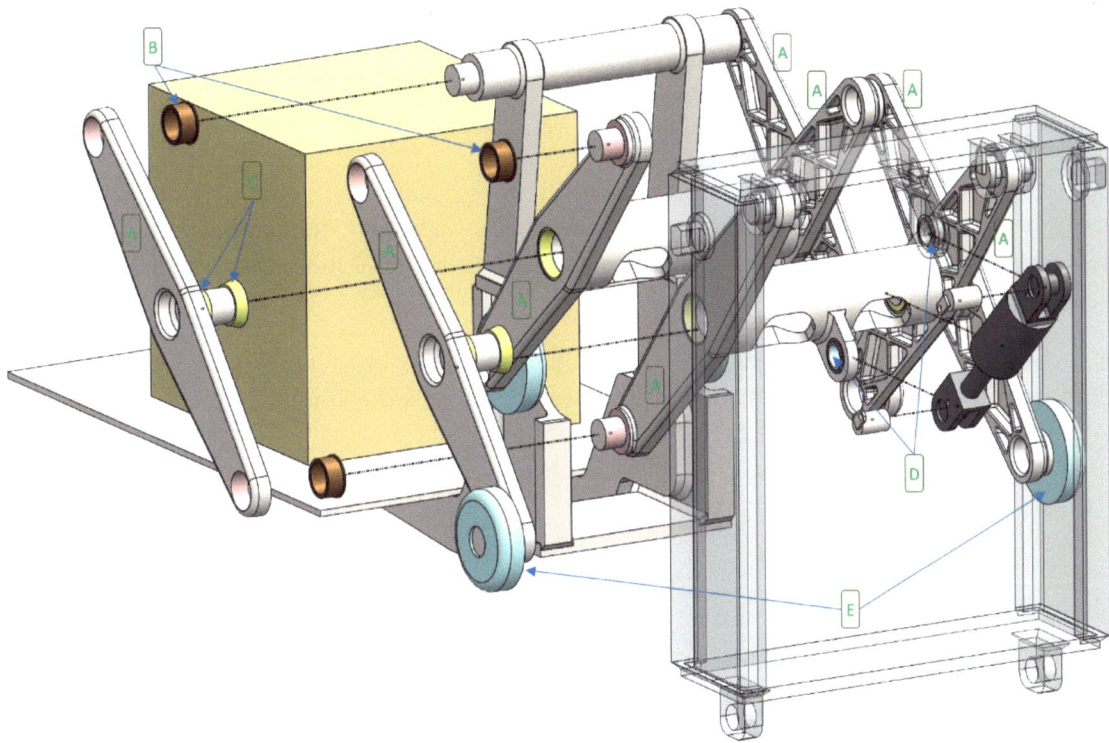

Partial exploded view of assembly

A total of six bronze bushings (B) are used. Where the arms cross pairs of tapered roller bearings (C} are installed, four pairs of inner and outer races. The reach cylinder is mounted by pins to a ball joint (D) on either end. The front and rear lower arm ends carry rollers (E) which ride in rails in the front and rear frames.

Tapered bearings

Each pair of tapered bearings could be handled as a single pin or bearing connector, if we were ready to assume that each is effective and local stresses were not of concern. For this assembly we will be more literal, but still keep things simple.

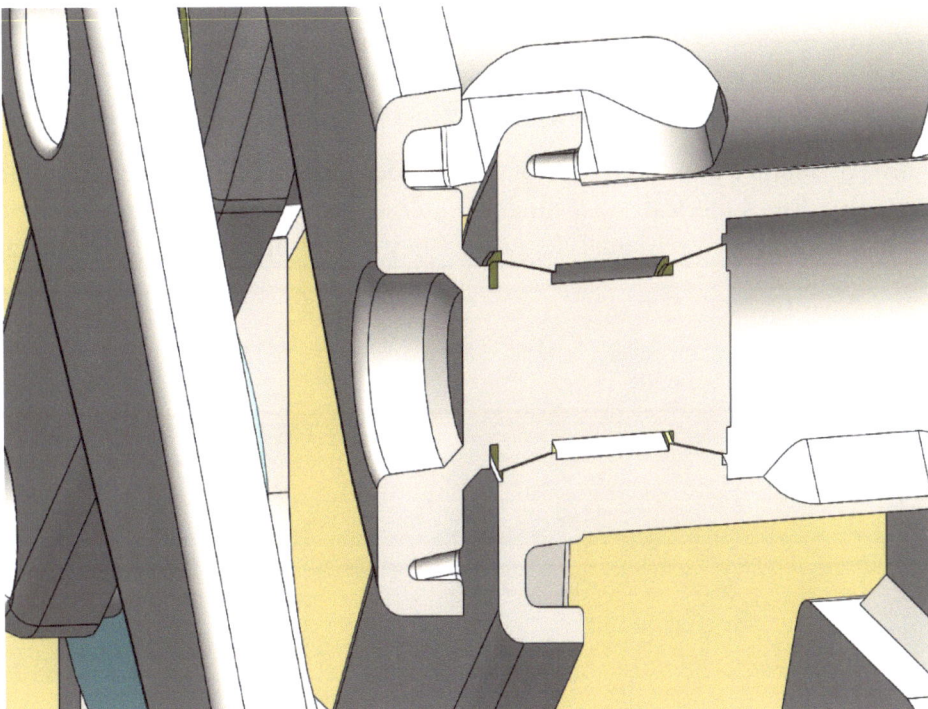

Section through left rear tapered bearing pair

The bearing races are modeled as solid parts of the main components, the inner races on the outside arm and the outer races on the central arm set. In reality two of the races must be installed from inside the tube (which is why it has large cutouts) and the races pressed in or retained in some other fashion. That detail is omitted here.

The faces of the races have a small clearance between them. No penetration contact conditions will be defined at each pair. This simulates a snug fit. If the installation would more firmly clamp the tapers tight to each other, that cannot be represented this way.

Another section through the middle arm-to-arm joint shows the bronze bushing being simply slipped between machined surfaces of each arm. Four surfaces of the bushing will be defined in contact, the inner and outer cylindrical faces and both flat sides of the small flange.

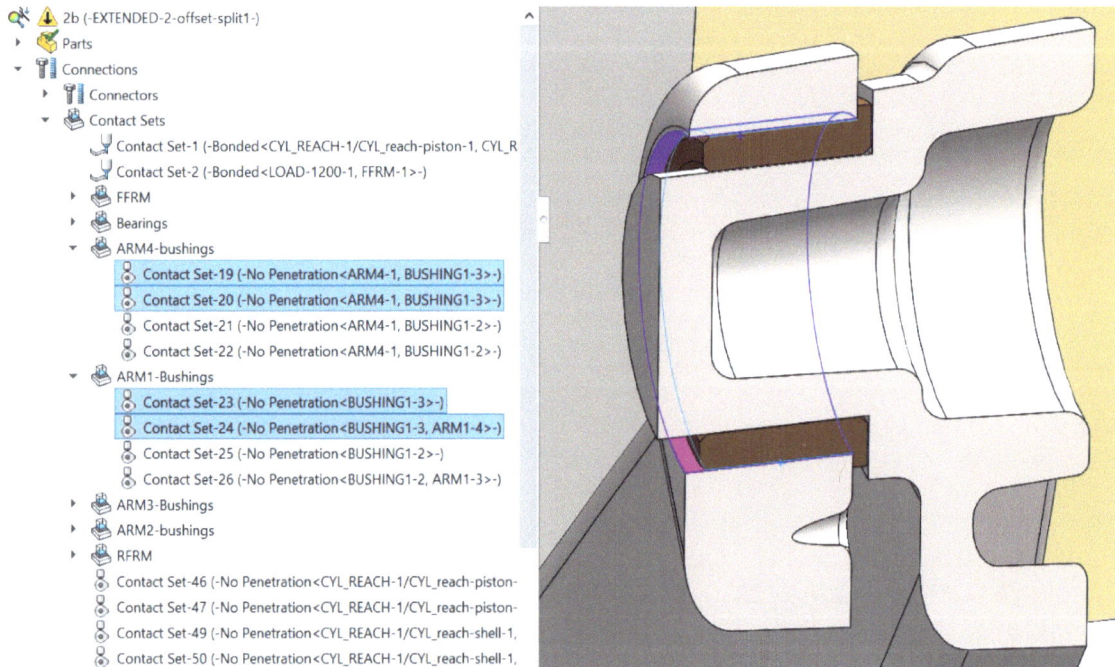

Section through upper middle left arm joint

There is no retainer on this bushed joint; the outer arm can slip off freely. The only thing resisting any lateral forces between the arms is the tapered bearings.

As noted, there are ball joints where both ends of the cylinder mount. These are modeled directly into the mating components. The spherical ball is modeled literally as a separate solid. The concave and convex spherical surfaces have a very small clearance. They can be simulated in no penetration contact and will behave as intended.

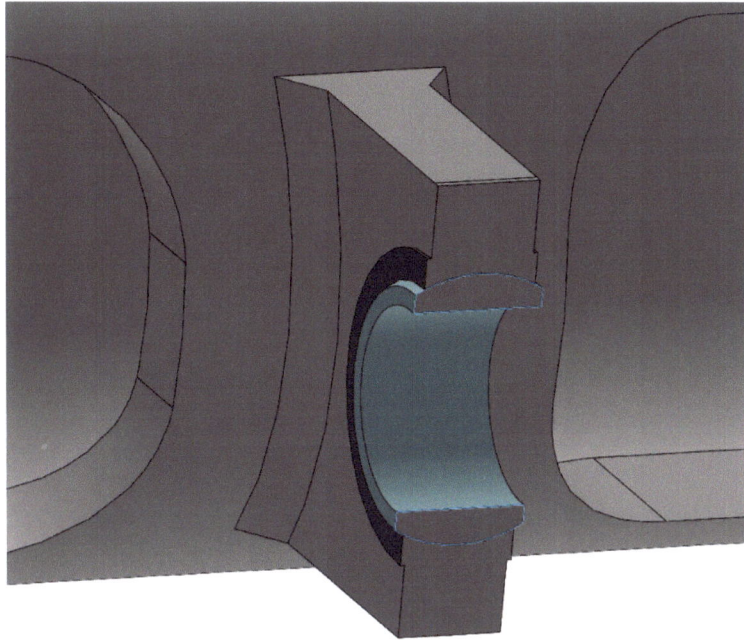

Section through rear center arm set at ball joint

At the rear of the assembly four pins with bolted tabs join the rear arms to the rear frame and locate both ends of the reach cylinder. For initial studies we will not consider the bolts through the tabs; each tab will be bonded to the near mating face.

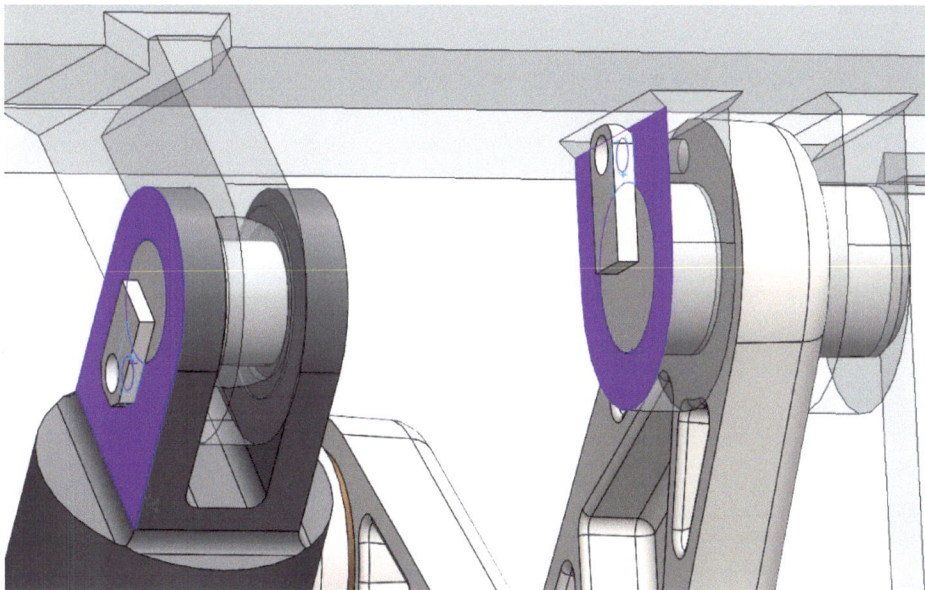

Pins with bolted tabs, bonded contact highlighted

Initial Setup

The reach mechanism is supposed to have a cargo capacity of 1200 pounds mass. This is a little more than one full cubic foot of solid gold. For the first studies we will model the load, place it at the center of the platform, and "bond" it in place.

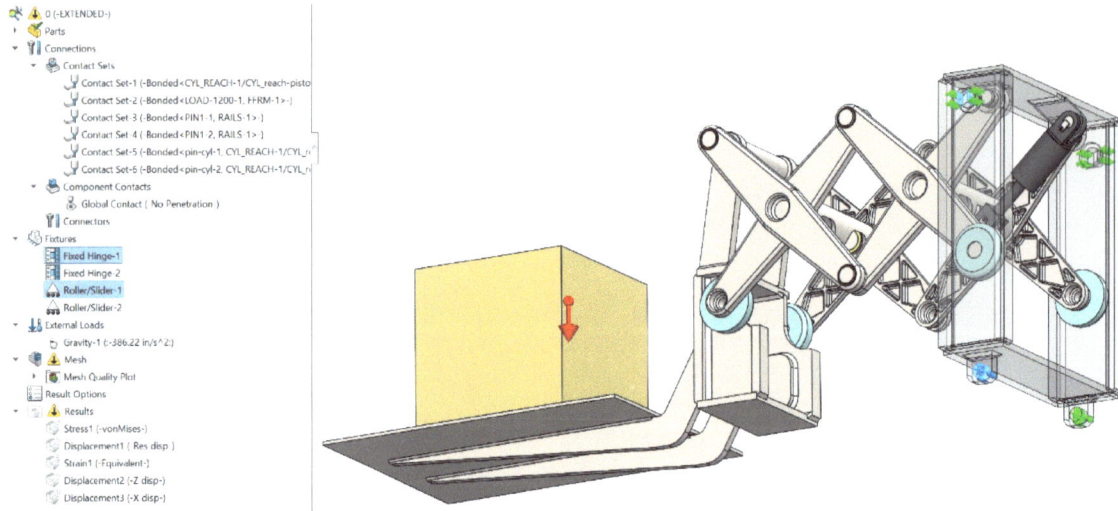

Initial simulation setup

The rear frame is constrained where it connects to the carousel; "fixed hinges" at the bottom where it is pinned to lifting cylinders and "roller slider" conditions higher up where tabs slide in tracks. Bonded contacts are made in the reach cylinder, under the load, and at the pin retainers as previously mentioned.

A gravity vector is defined straight downward at 1 g. There are *no* synthetic applied forces. This is a very important point. We have eliminated a manual step and source of assumption and (too frequently) error. By modeling the load literally, we let the solver do all the ciphering and can focus on a setup that 'looks' right.

Again, since the model looks to have all parts in close clearance as we need, we can start with a single "global no penetration contact"

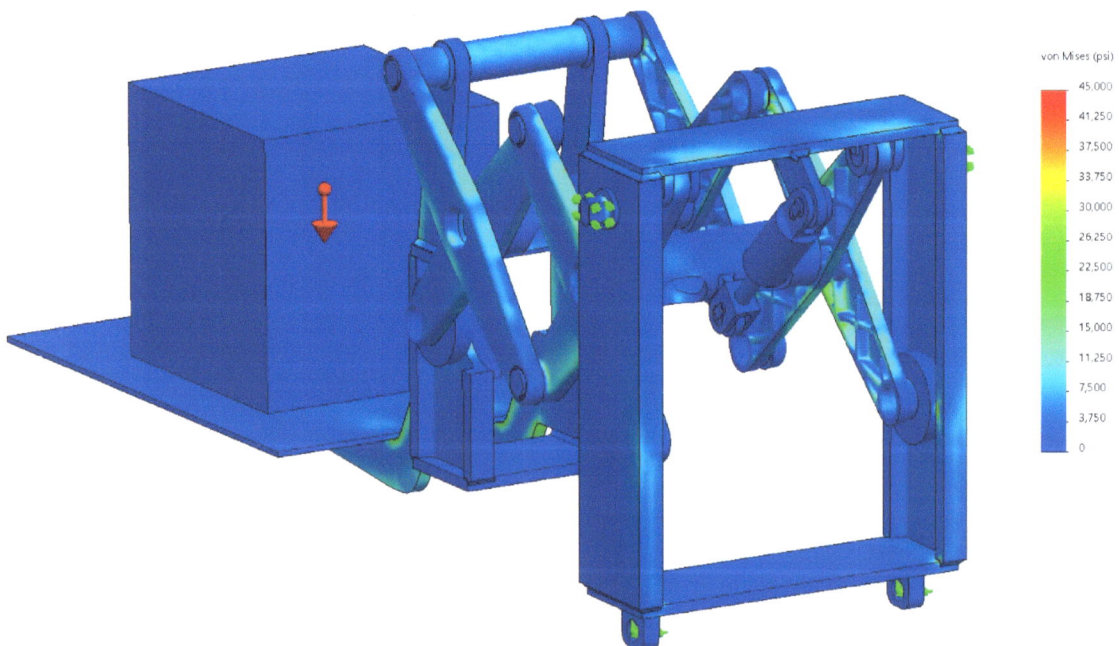

Initial result

We also make a run with the mechanism retracted.

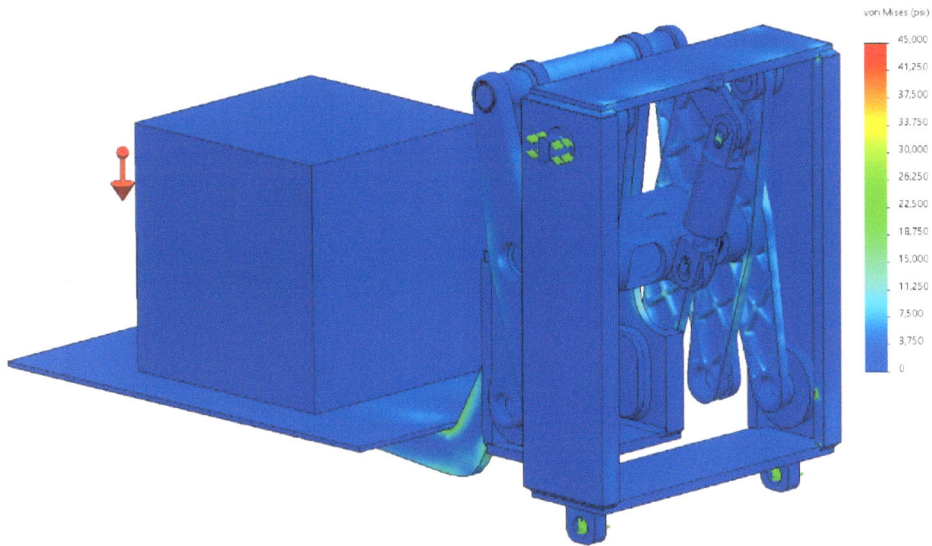

Initial retracted result

Stresses are pretty tame in both cases, especially with the load retracted. But this is for a static condition, with the load centered and nothing happening. We must look at what happens when the carousel moves.

Inertial Loading

For starters we increase gravity to 1.15 g, as if the lift cylinders were accelerating the machine upward. The load is moved off to one side, as not every placement will be perfect.

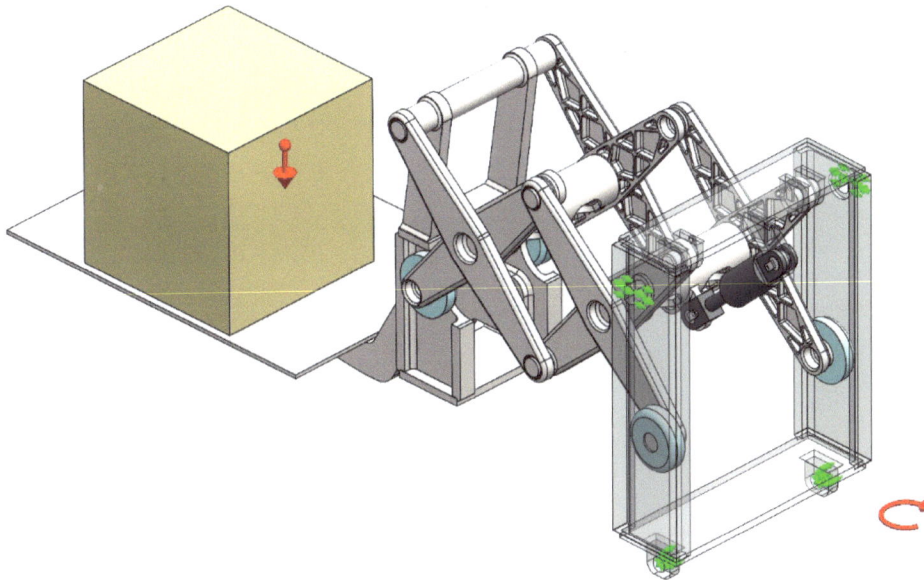

Setup with motion

Then a rotational acceleration is added.

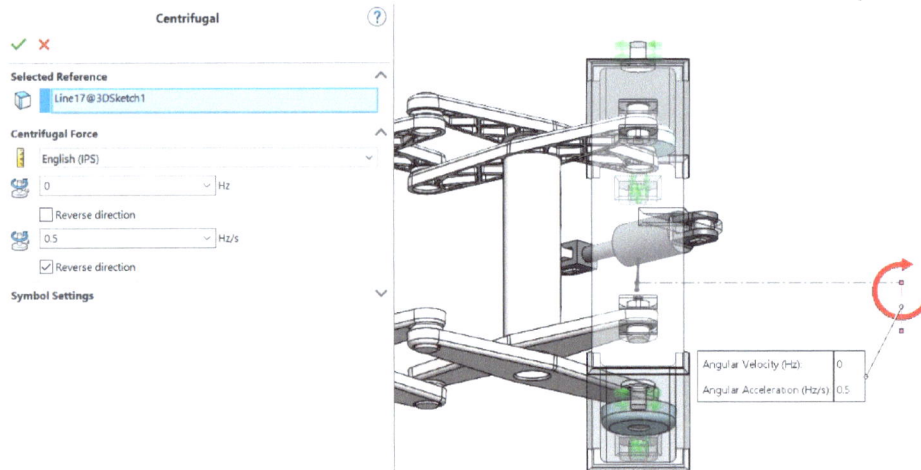

Definition of rotational acceleration

The carousel designers tell us their part of the machine is capable of spinning up at a rate equivalent to 0.5 Hz per second. (The other options are radians per second and rpm per second; we do not expect other designers to offer up numbers in any of those ways.) We are still doing a *static* analysis. These accelerations are applied as if they are constant, or applied for a long enough time for the machine to get over any bouncing and jiggling from the disturbance. We will not capture the true dynamic overshoot seen when an object is disturbed from rest. We will however see it respond and move very realistically to real overloads.

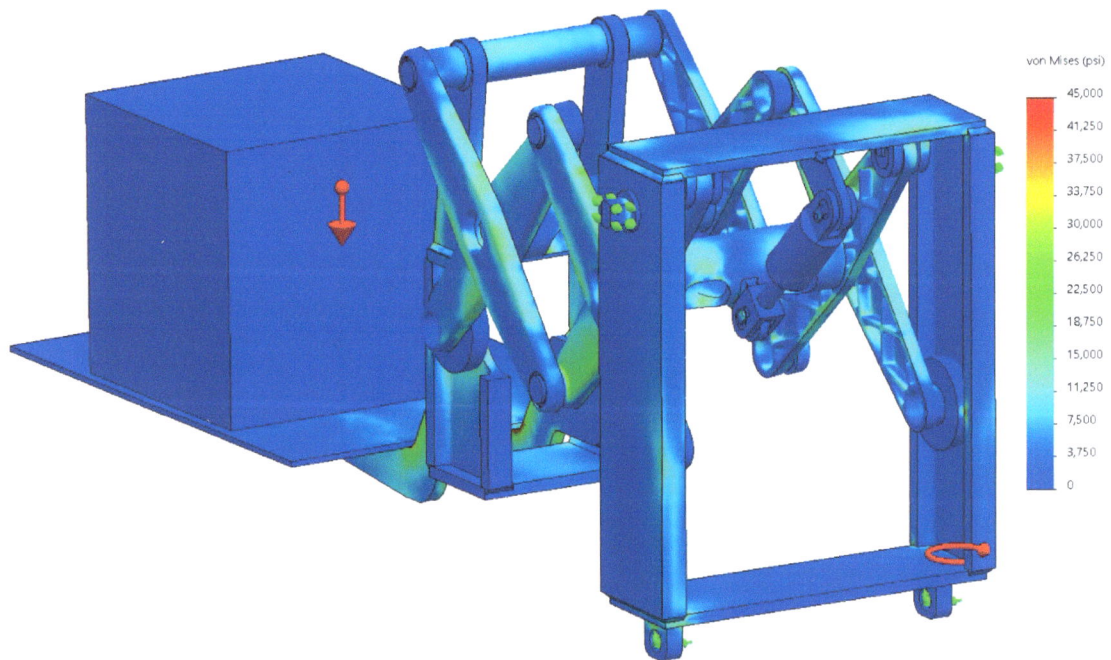

Result with added accelerations

Magnified displacement plots show the inertia of the load resisting the attempt at rotation.

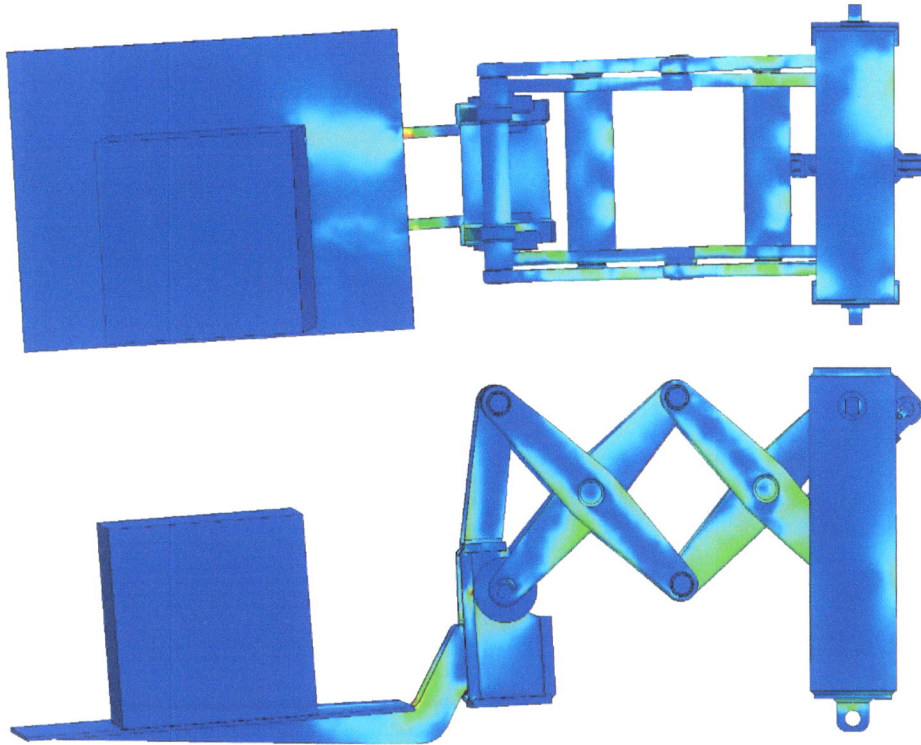

Result with added accelerations, 5X displacement

We can look more closely at stresses, but first note that so far we have been working with a first-pass coarse mesh. It has served well to establish that we have a good working model, but it can be improved. There is also at least one very high aspect ratio (1314) element, but we ignored mesh quality checks at the start.

Mesh Details	
Study name	2 (-EXTENDED-offset
Mesh type	Solid Mesh
Mesher Used	Curvature-based mesh
Jacobian points for High quality mesh	4 points
Max Element Size	30 mm
Min Element Size	6 mm
Mesh quality	High
Total nodes	409915
Total elements	237486
Maximum Aspect Ratio	1,314.3
Percentage of elements with Aspect Ratio < 3	77.8
Percentage of elements with Aspect Ratio > 10	1.24
Percentage of distorted elements	0
Number of distorted elements	0
Remesh failed parts with incompatible mesh	Off
Time to complete mesh(hh:mm:ss)	00:00:13
Computer name	ARMISTICE

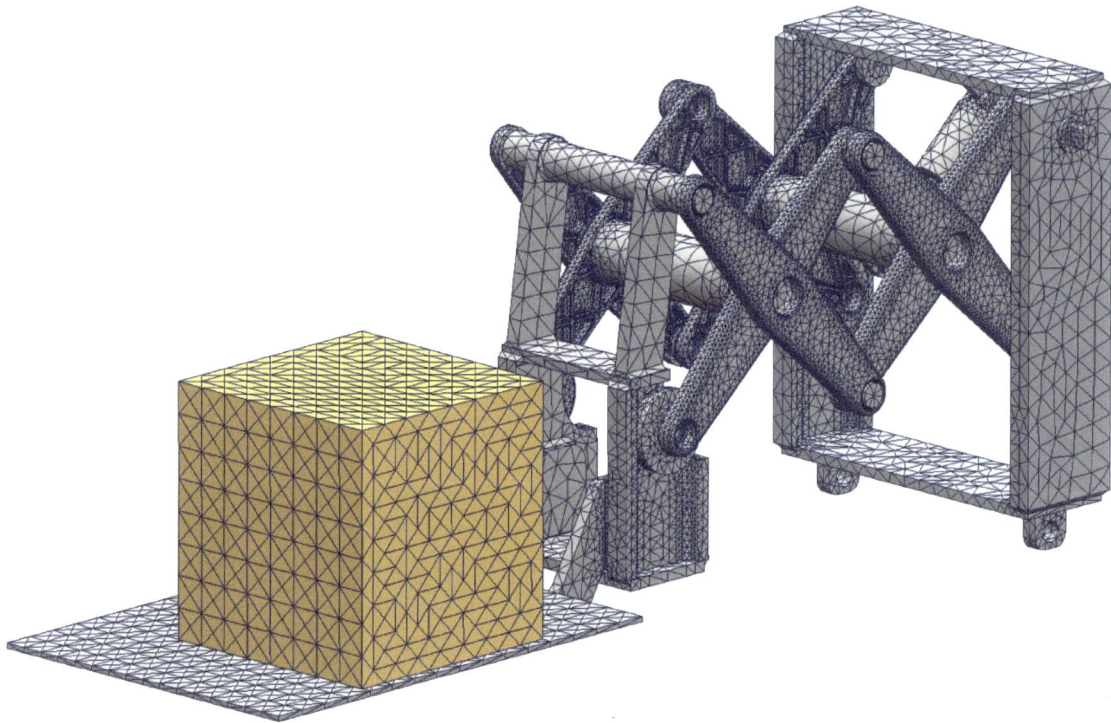

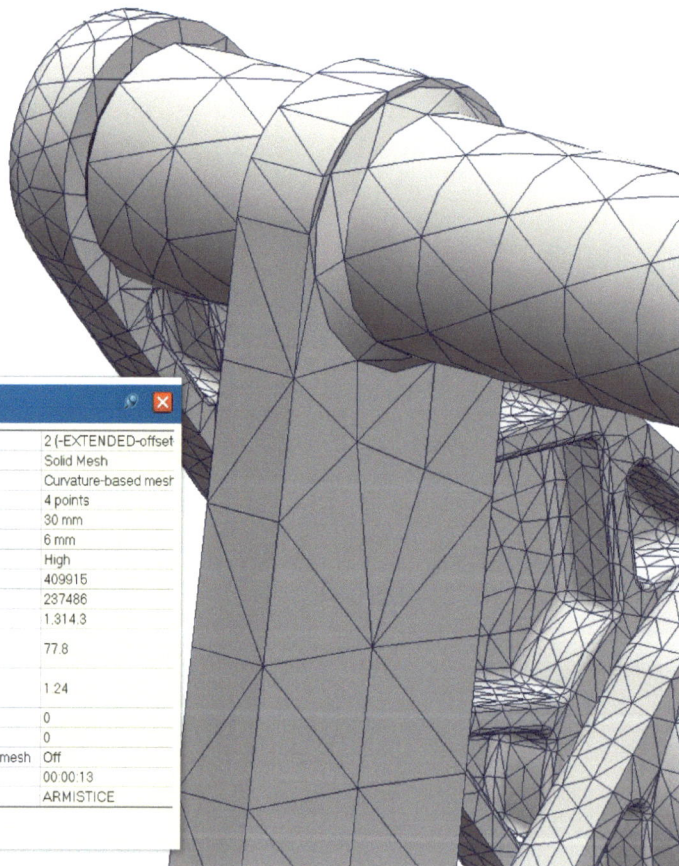

Initial mesh, overview and detail

The curvature-based mesher has helped us with refinements on the small radii of the arm castings. But on the wrought parts mesh sections are very coarse, including one-element-thick portions where bending is expected. Locations where contact is expected are also refined.

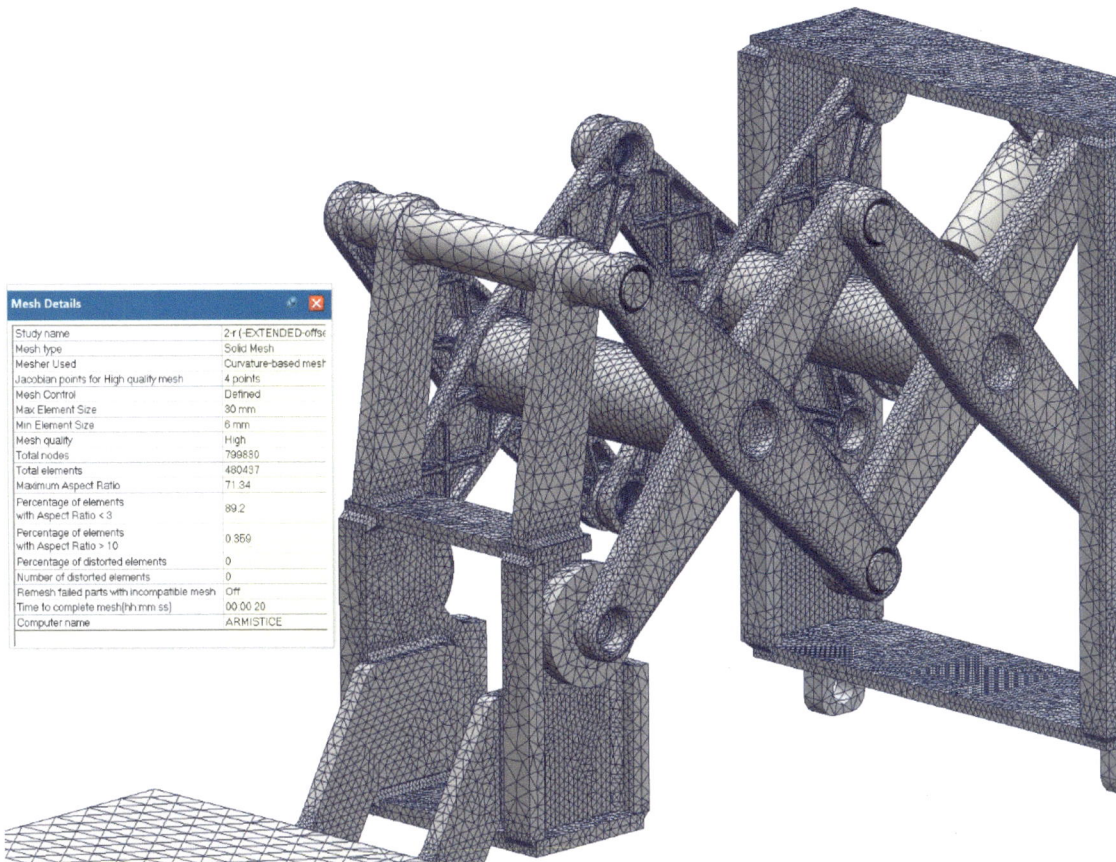

Mesh Details	
Study name	2-r (-EXTENDED-offse
Mesh type	Solid Mesh
Mesher Used	Curvature-based mesh
Jacobian points for High quality mesh	4 points
Mesh Control	Defined
Max Element Size	30 mm
Min Element Size	6 mm
Mesh quality	High
Total nodes	799880
Total elements	480437
Maximum Aspect Ratio	71.34
Percentage of elements with Aspect Ratio < 3	89.2
Percentage of elements with Aspect Ratio > 10	0.359
Percentage of distorted elements	0
Number of distorted elements	0
Remesh failed parts with incompatible mesh	Off
Time to complete mesh(hh:mm:ss)	00:00:20
Computer name	ARMISTICE

Mesh after manual refinement

We run again with the new mesh and sharper details show up in the stress pattern.

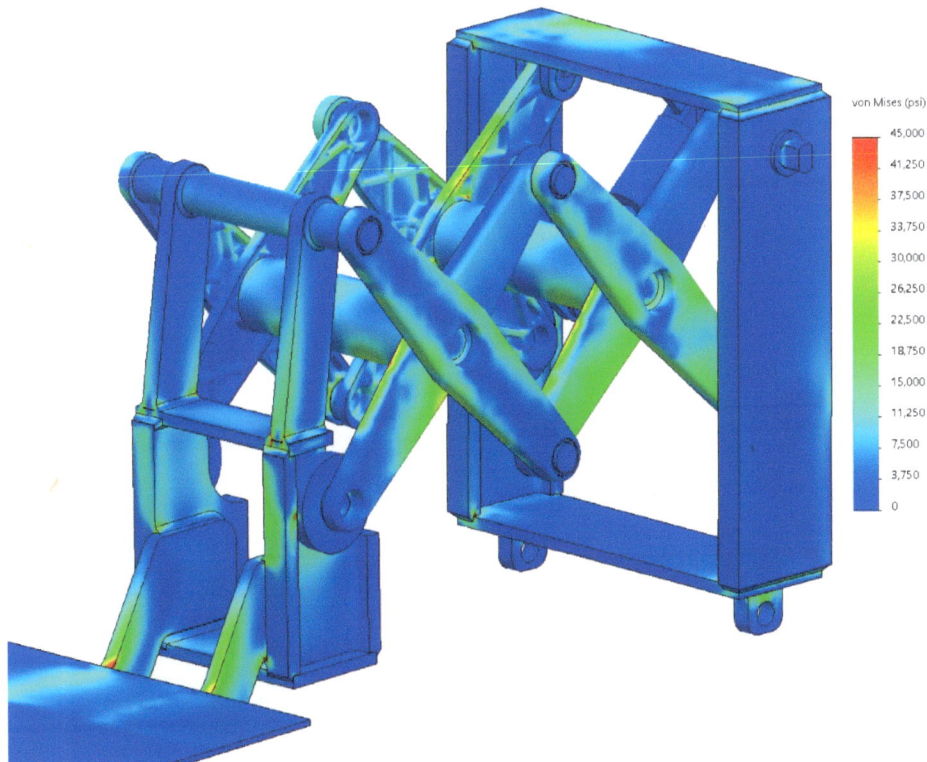

Result with refined mesh

We want to get in and look at the stress hot spots more closely. But we also want to pull out contact forces. Up to now we have run global no-penetration contact. In this condition we can never be sure where the solver put all the forces, such as all on mating flat faces or perhaps bled over onto adjacent small fillets. It also takes longer and uses more resources when the solver must consider every surface node possibly interacting with every other surface. We want to take control.

Splits for Contact

This process starts by splitting surfaces so that we can identify discrete mating pairs. The rollers are split so that front and back halves of the rolling face are separated. The rails they roll in are also split so only a short section of each is included in the contact sets.

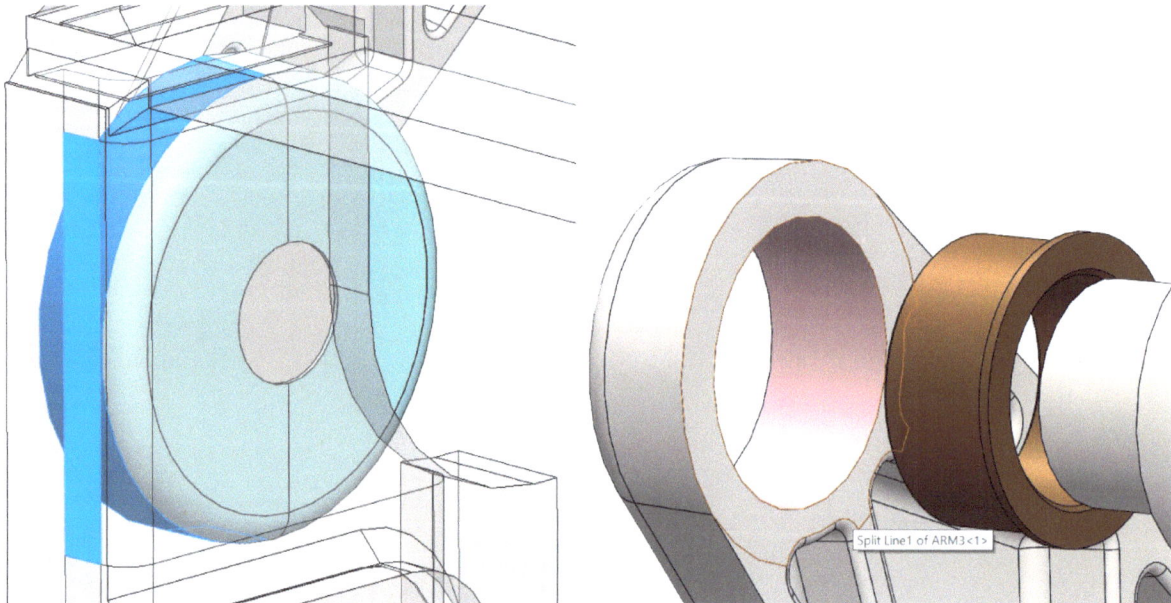

Surface splits for contact sets

The end of the inside faces of each arm are also split around the bushing locations, to make a single surface which will be defined in contact with the flange of the bushing. Contact pairs are then manually defined for each roller (front, back, and side), bushing (2 cylindrical and 2 flat surfaces), tapered bearing, ball joint, and pin surface. Over 60 contact sets are defined.

While we are improving details the pin retaining bolts are defined and the tabs put into no-penetration contact.

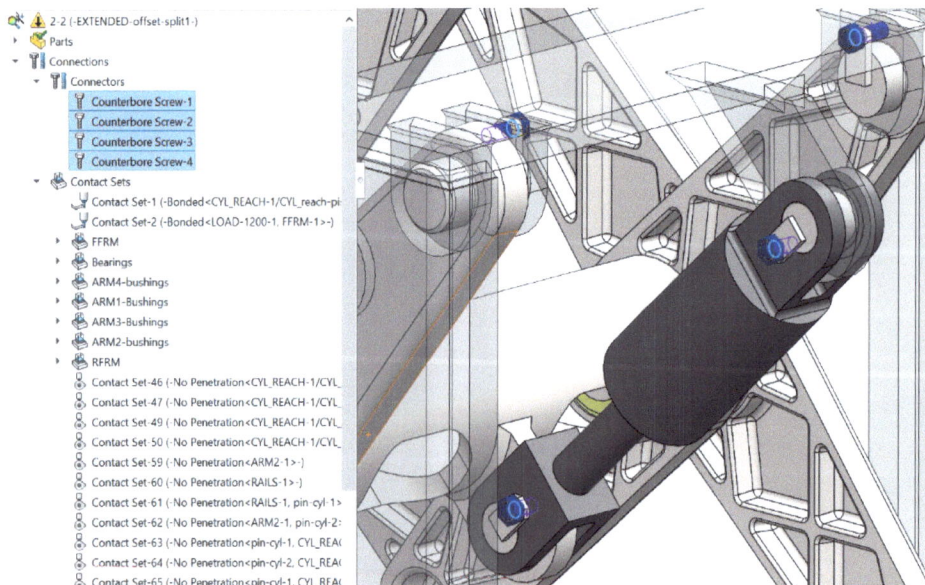

Pin retaining bolts defined

With all the literal detail we can think of put into the model, and a good mesh laid on, it we run again.

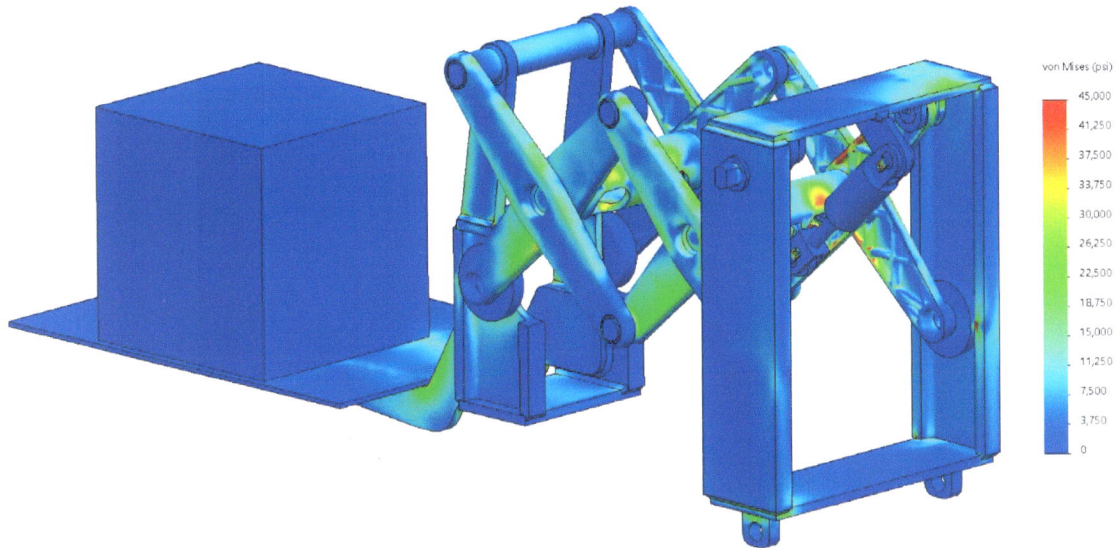

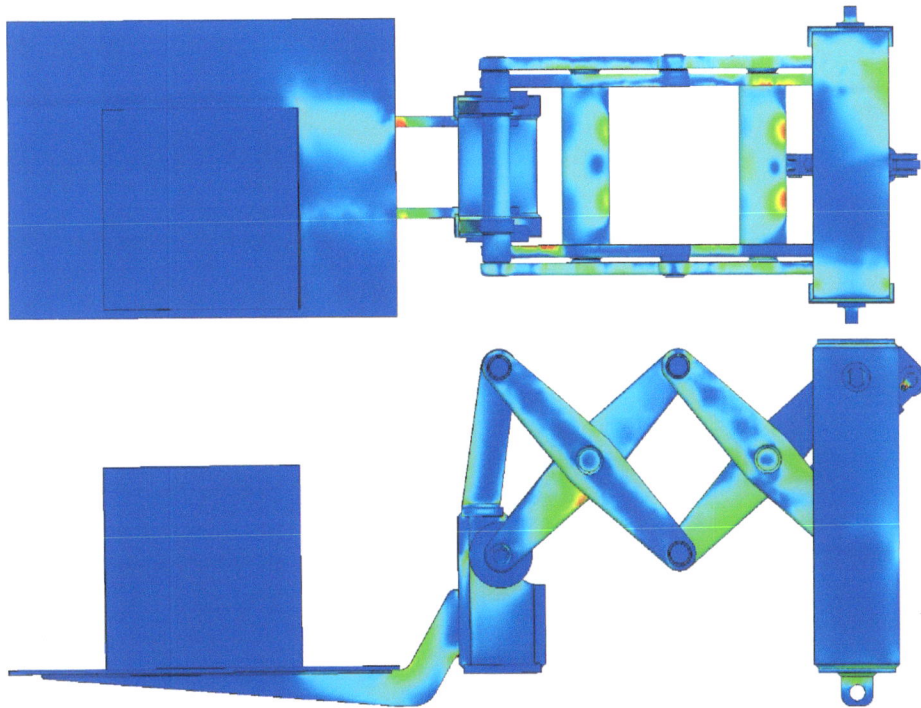

Result with manual contact

Now we can start to look closely at different areas on the machine. The front frame shows some hot spots near weld ends, which are expected and not thought to be a problem for weld filler material. One of the platform arms is a bit warm, but not badly.

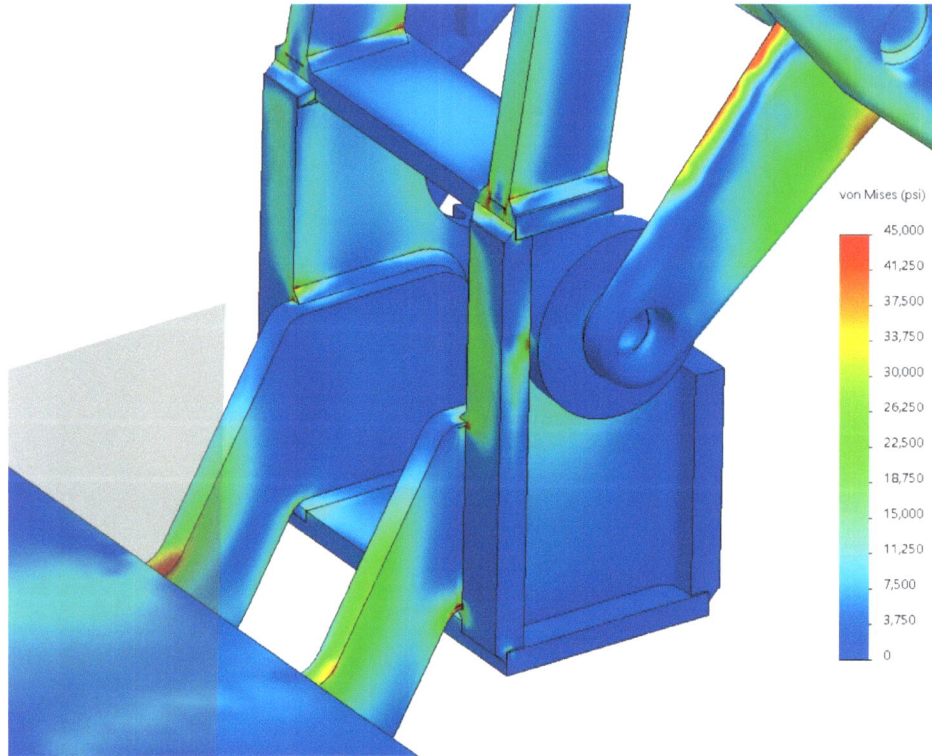

Front frame detail

What really grabs our attention is the front lower arm. (Following views will have various parts hidden.)

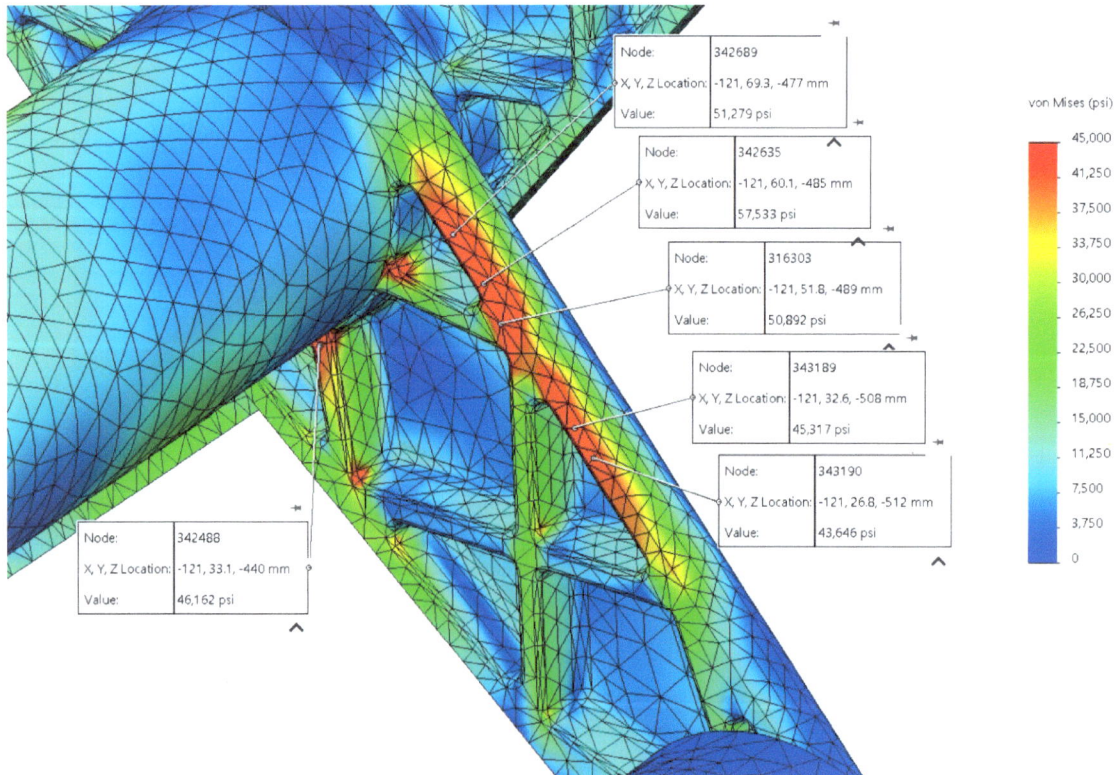

Node:	342689
X, Y, Z Location:	-121, 69.3, -477 mm
Value:	51,279 psi

Node:	342635
X, Y, Z Location:	-121, 60.1, -485 mm
Value:	57,533 psi

Node:	316303
X, Y, Z Location:	-121, 51.8, -489 mm
Value:	50,892 psi

Node:	343189
X, Y, Z Location:	-121, 32.6, -508 mm
Value:	45,317 psi

Node:	343190
X, Y, Z Location:	-121, 26.8, -512 mm
Value:	43,646 psi

Node:	342488
X, Y, Z Location:	-121, 33.1, -440 mm
Value:	46,162 psi

Probed results on front lower arm

These stresses come from forces which we can check on the roller.

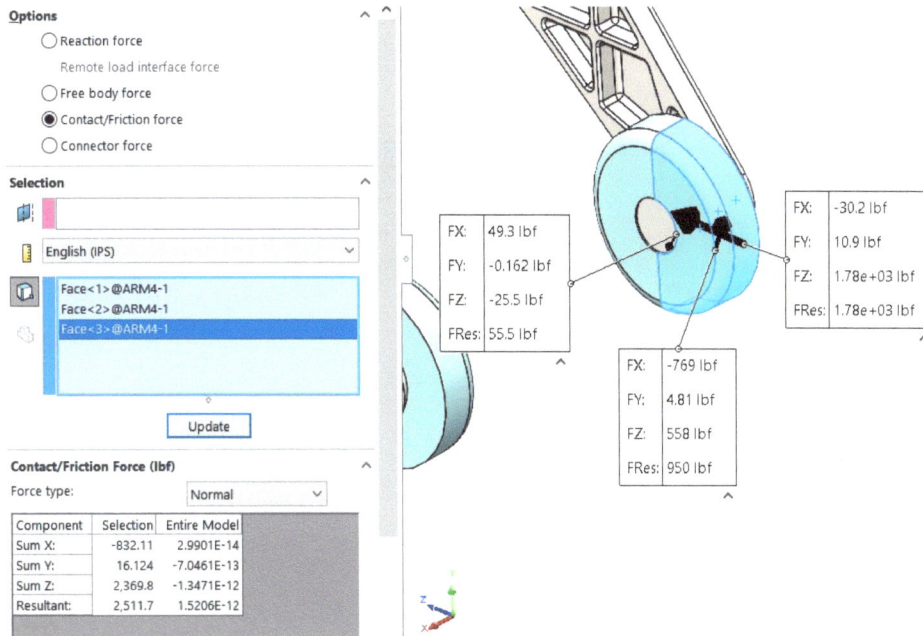

Options
- Reaction force
  - Remote load interface force
- Free body force
- Contact/Friction force (selected)
- Connector force

Selection

English (IPS)

Face<1>@ARM4-1
Face<2>@ARM4-1
Face<3>@ARM4-1

Update

Contact/Friction Force (lbf)

Force type: Normal

Component	Selection	Entire Model
Sum X:	-832.11	2.9901E-14
Sum Y:	16.124	-7.0461E-13
Sum Z:	2,369.8	-1.3471E-12
Resultant:	2,511.7	1.5206E-12

FX:	49.3 lbf
FY:	-0.162 lbf
FZ:	-25.5 lbf
FRes:	55.5 lbf

FX:	-30.2 lbf
FY:	10.9 lbf
FZ:	1.78e+03 lbf
FRes:	1.78e+03 lbf

FX:	-769 lbf
FY:	4.81 lbf
FZ:	558 lbf
FRes:	950 lbf

Forces on roller

Because there is no discrete lateral load bearing component we had put all surfaces of the roller in contact with the rail and let the solver resolve the balance. The exact distribution of forces will be very sensitive to model clearances, curvature, and local mesh refinement.

Model Change

The model is changed to thicken ribs on all the arms. We take advantage of the parametric model to update geometry in a way that requires no changes to the simulation setup.

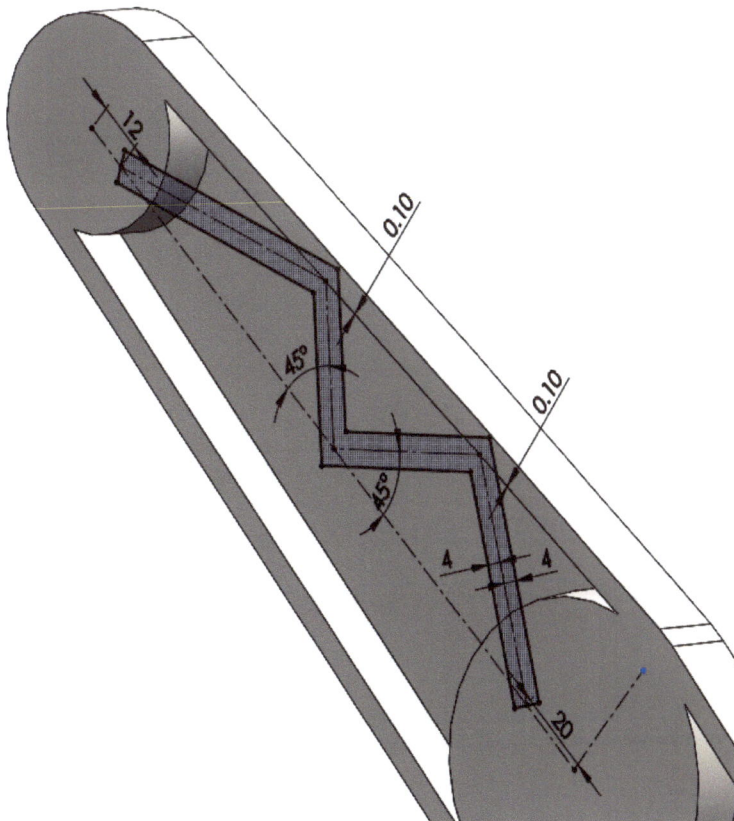

Model dimension updates

All surface IDs remain the same and the splits done for simulation rebuild in the new configuration without edits.

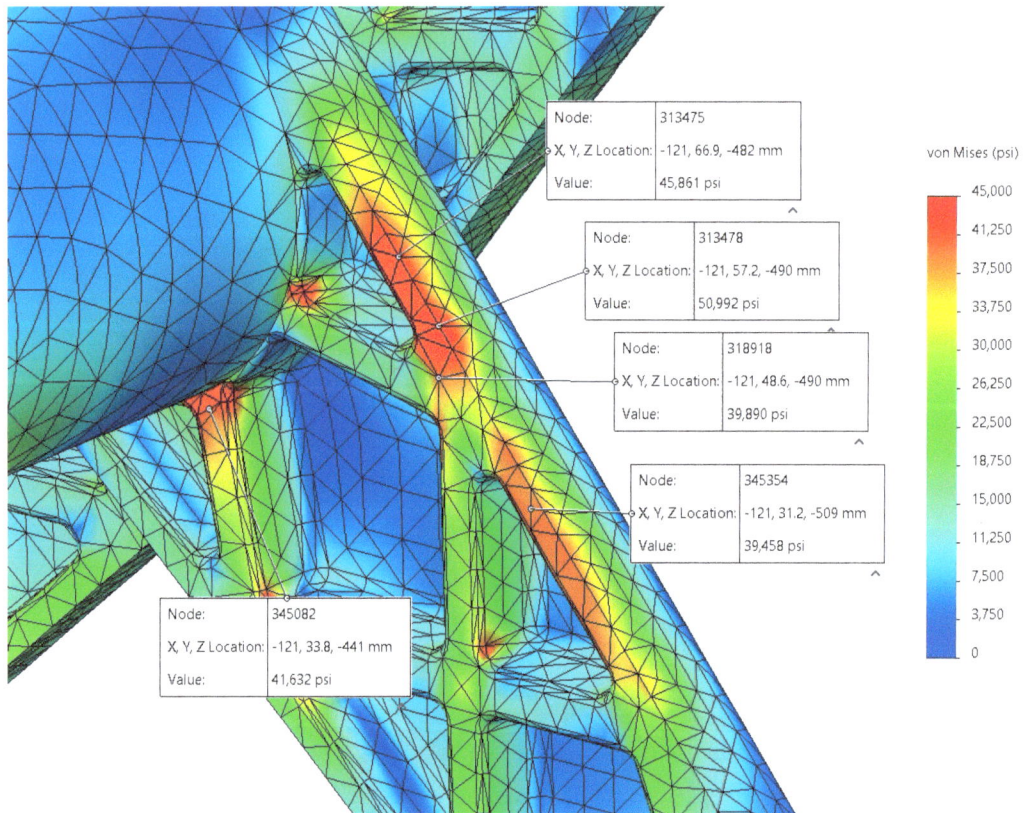

Result with thickened ribs

The stress peaks can be discussed with the casting engineer. With the right alloy they may be acceptable.

Additional discussions can be had about the wrought tubes, bearings, and other components based on detailed stress and force results.

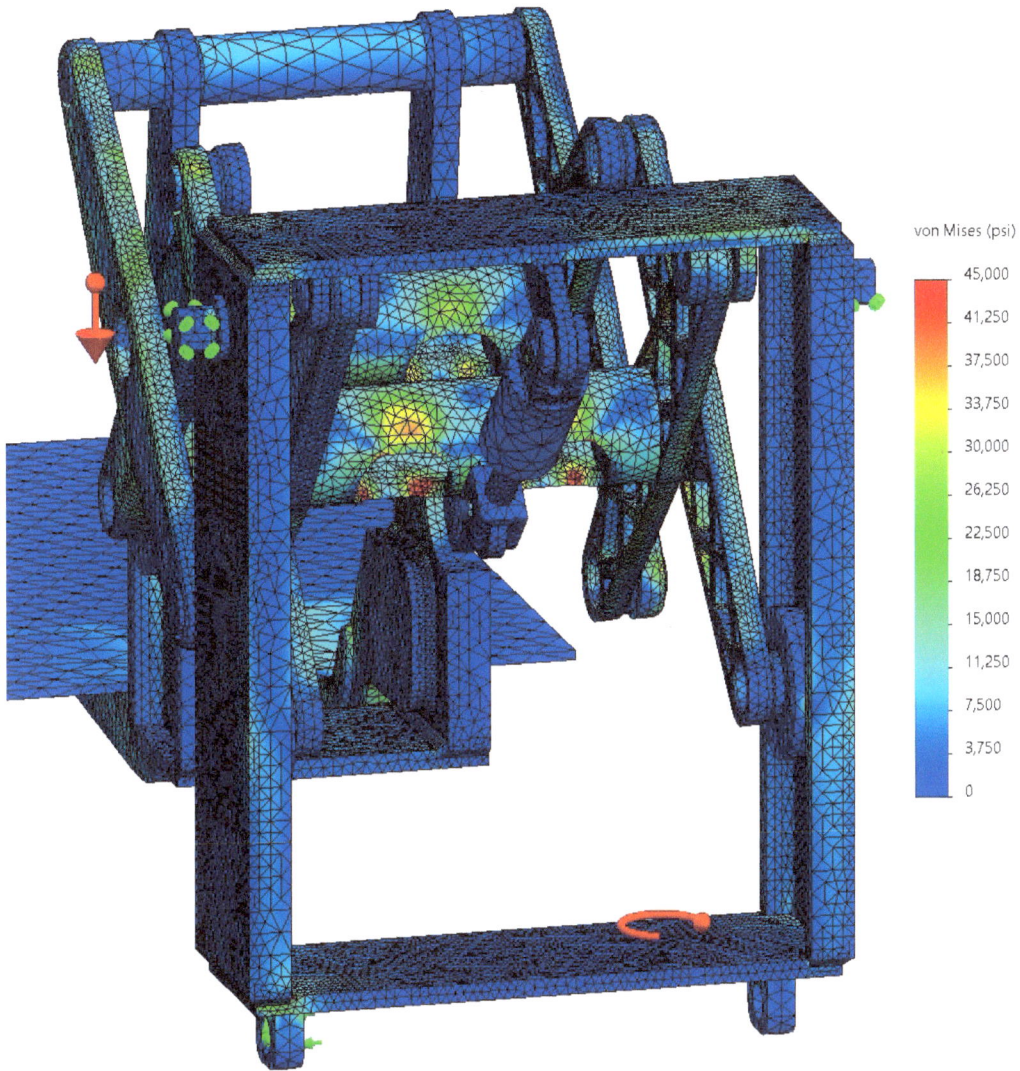

Overall result with mesh

Retracted Model

With some planning the retracted version of the model can be put into a copied simulation study with no edits to the setup.

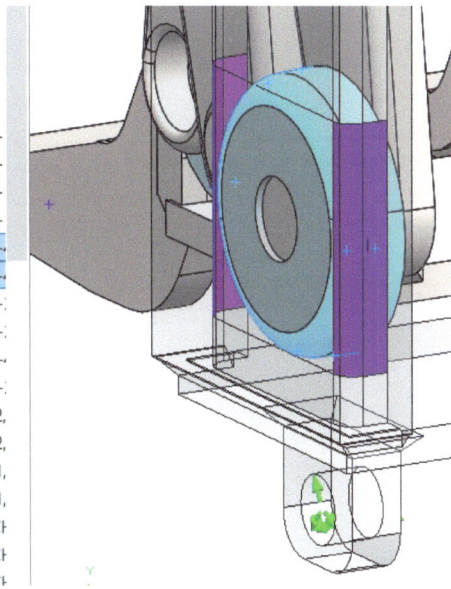

Contact setup on rear roller

Splits on the front and rear rails were defined by a sketch with configurable dimensions. If different part configurations are used for extended and retracted assembly configurations the contact definitions update as intended.

All result data types can be generated on the retracted model.

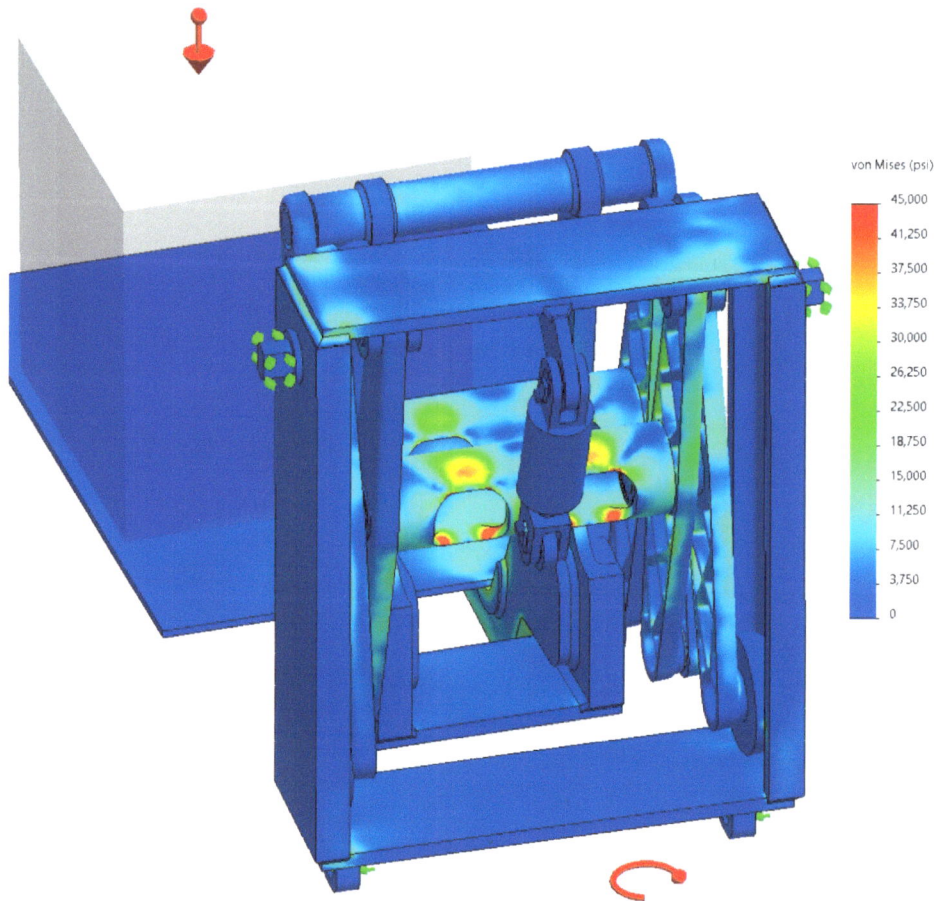

Result on retracted model

Bearing Connector

A connector type we have not used yet is the "bearing". This connector applies to a pair of internal and external cylindrical surfaces. The ball models in our cylinder clevises are made merged solids.

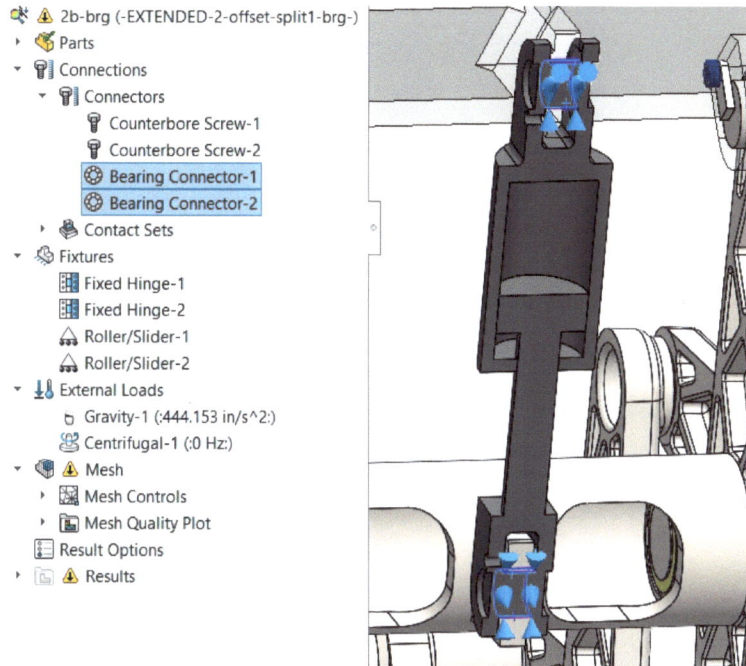

Bearing connector locations

To use the bearing connector the cylinders selected should be centered, coaxial, and about the same length. The "Allow self-alignment" option lets the bearing behave as a ball joint (if this option is not needed, the pin connector is probably a better choice).

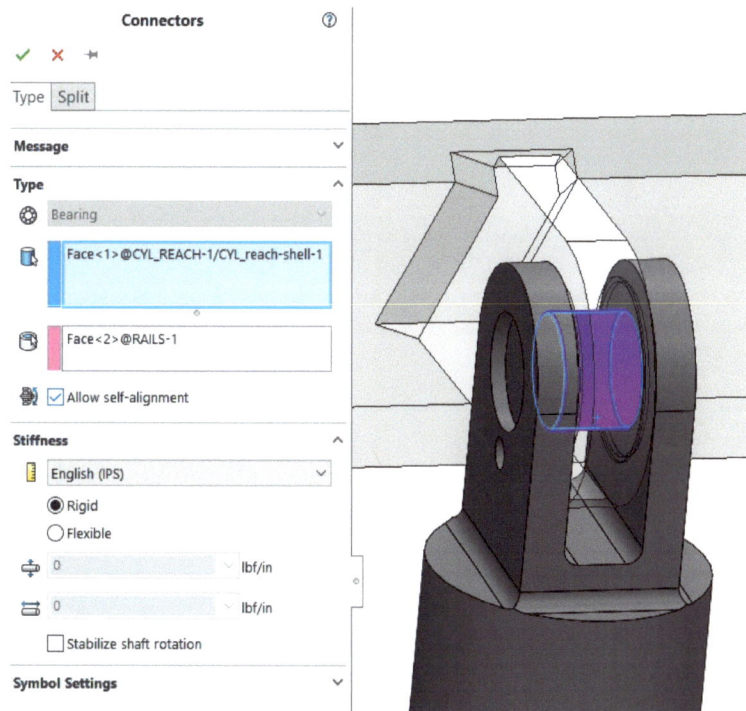

Bearing connector definition

The bearing connector does distribute stress through the joint.

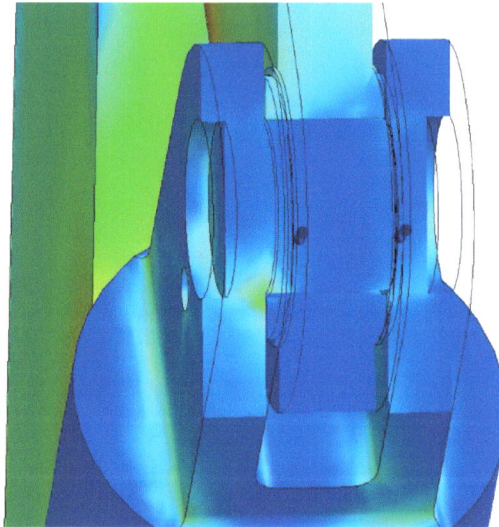
Result through bearing joint

Run time improves substantially using the bearing connector instead of the balls in full contact. It may be worth adjusting ball joint models into simpler solids in some circumstances.

Reach Obstruction

It is possible that an obstruction will interfere with motion of the carousel or reach mechanism. We can simulate any such situation by simply modeling it literally. For the last example here we will park a fork lift such that the rear corner hits the load platform near one edge.

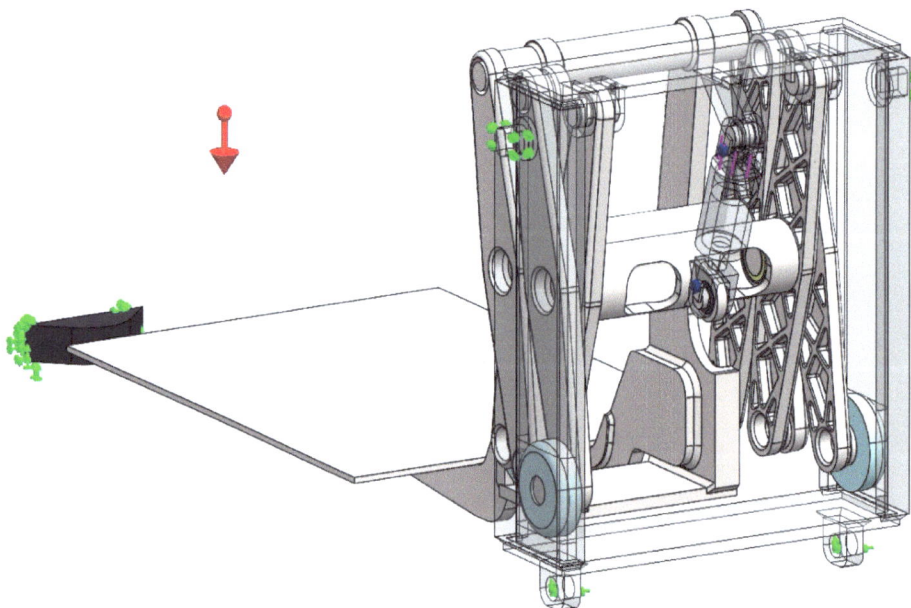
Setup with reach obstruction

A small part of the lift truck counterweight is fixed in the model close to the platform. A contact condition is defined between the platform and truck.

Loading now comes from inside the machine. The reach cylinder internal pin connector is made loose and expected maximum hydraulic force applied to piston and shell faces.

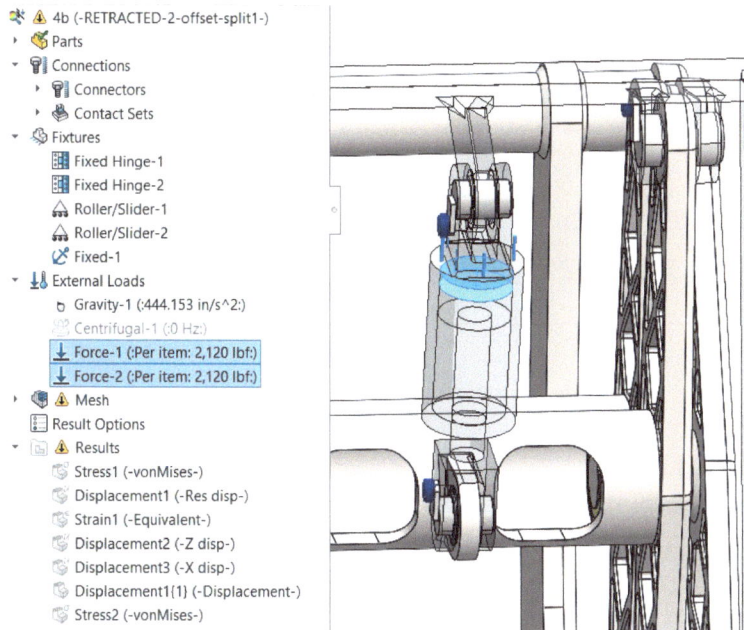

Forces applied to cylinder faces

The study runs without trouble and we get a new stress pattern. The offset obstruction puts asymmetric stress in the arm sets and through into the driving cylinder.

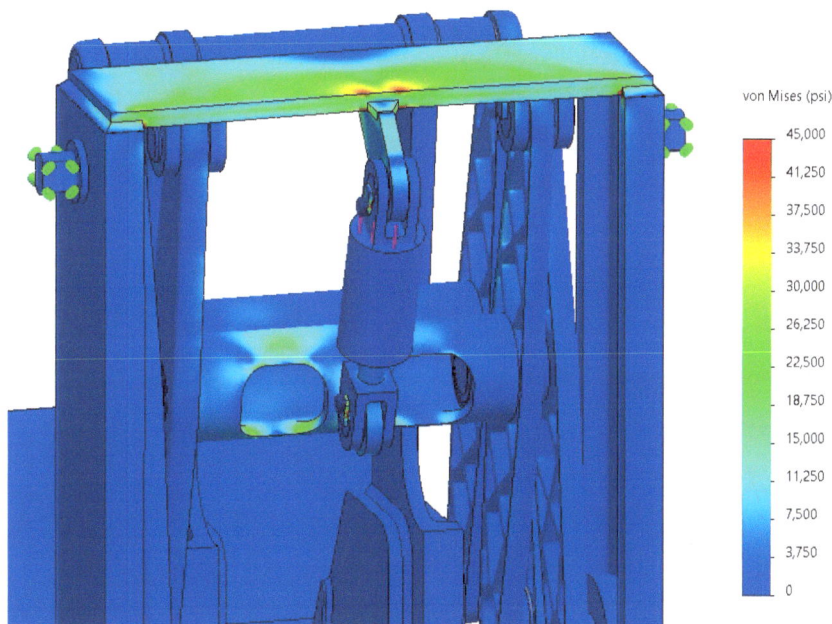

Result with obstructed reach

With full force in the hydraulic cylinder the thin top plate of the rear frame shows its low stiffness. Another design iteration may be needed.

				grav	rot	contact		Elem count	Run time
0	normal, ext	extended	1200	1	0	global			
1	normal, retracted	retracted	1200	1	0	global		100%	100%
2	offset, moving	extended	1200	1.15	0.5	global		100%	88%
2-r	refined mesh	extended	1200	1.15	0.5	global		202%	610%
2-2	refined mesh, manual contact	extended	1200	1.15	0.5	manual, surf-surf		209%	901%
2b	thicker arm ribs	extended	1200	1.15	0.5	manual, surf-surf		213%	757%
2b-brg	cylinder joints to bearing connectors	extended	1200	1.15	0.5	manual, surf-surf		212%	572%
2b-nobolts	pin bolts to bonded contacts	extended	1200	1.15	0.5	manual, surf-surf		213%	384%
3b	as 2b, retracted	retracted	1200	1.15	0.5	manual, surf-surf		215%	1016%
4b	retracted, pushing on obstruction	retracted	0	1	-	manual, surf-surf		211%	1111%

Tabulation of runs

The runs performed on this machine are tabulated above. The results follow a familiar pattern. Models with more contact and bolted joints take longer to run. The only way to be sure about what level of detail is necessary on a new problem is to try it all ways. We like to start simple and add complication in progressive evolutions.

Interesting to note is that for the same set of parts the retracted runs took longer. In physical reality the retracted shape of these machines is less stable; the reach mechanism has less advantage and side sway is easier to introduce. This reduced real stability comes through in the math of trying to get a static solve in simulation.

Study 8: Lift Mast

Overview
A fork lift mast is to be simulated under load in different static and dynamic conditions.

Objective
Stresses in various weldments are needed. Also desired are force data in a variety of cylinders, chains, sheaves, and rollers.

Challenges
The assembly is fundamentally unstable; each weldment hangs on chains, located by multiple rolling and sliding contacts. We must simulate chains over sheaves. Paired hydraulic cylinders add a new complication. The high aspect ratios of the rails can drive up the mesh size.

Three Stage Free-Lift Mast
The three stage lift mast is designed for a 3000 pound capacity. Total lift is 208 inches (5.3 m).

Three stage mast – down, full free lift, maximum extension

A pair of hydraulic cylinders in the rear, the "main lift", push the rail weldments apart giving most of the lift. The mast has a single forward cylinder providing "free lift" of the fork carriage without the rail weldments moving. This accounts for 68 inches of the lift.

Free lift cylinder

A top view overlaid with a root design sketch shows how the rails and rollers are aligned. Large rollers on the middle and outer rail weldments take up forward bending forces. Lateral forces on those rollers are carried through to a short ledge on each rail. The carriage runs in the inner weldment on smaller rollers with separate adjustable pucks to take lateral forces.

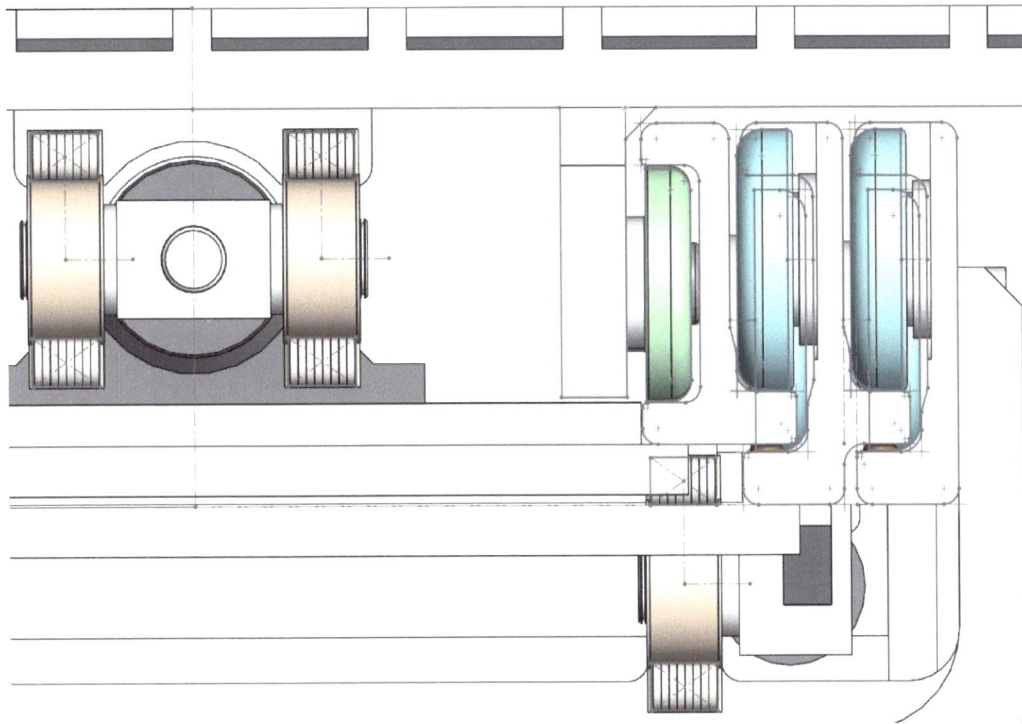

Top view of rail set.

Small wear plugs in the rear flange of the middle and outer weldments hold the weldments against possible rearward lean. All rollers and sliders are mated in the model to 0.5 mm clearance against the intended running surface.

At full extension the rail weldments are lifted to a twenty inch (508 mm) overlap. The 99 mm main rollers are attached 50 mm in from the rail ends, so primary roller forces will be spread on a 408 mm moment arm.

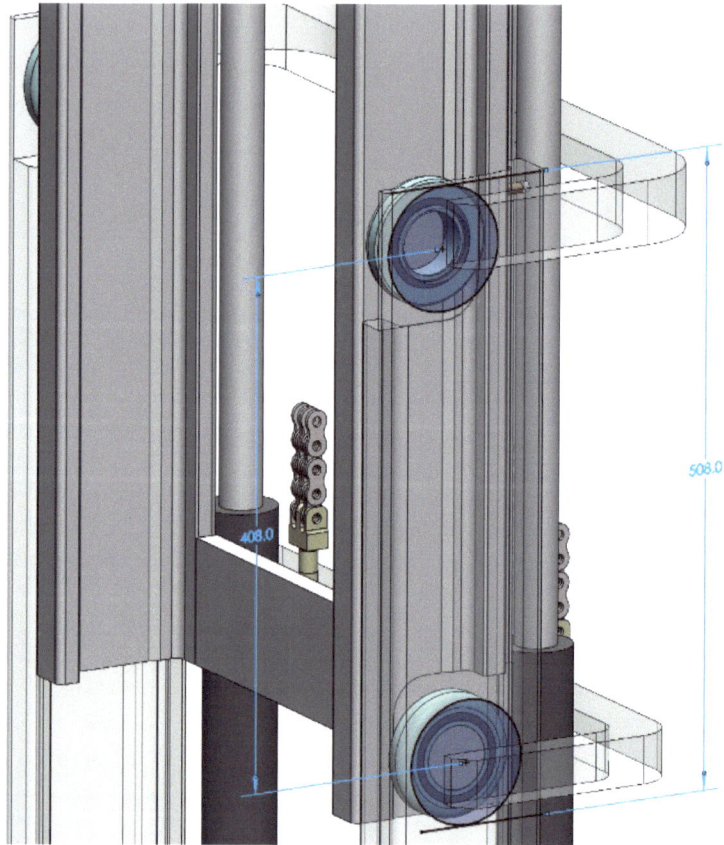

Rail and roller overlap

Front Chain Set

The free lift cylinder raises the four-roller fork carriage by a redundant pair of leaf chains. The end of each chain is pinned to an anchor bolt which passes through blocks on either end to bear on the reverse side of each block by a nut.

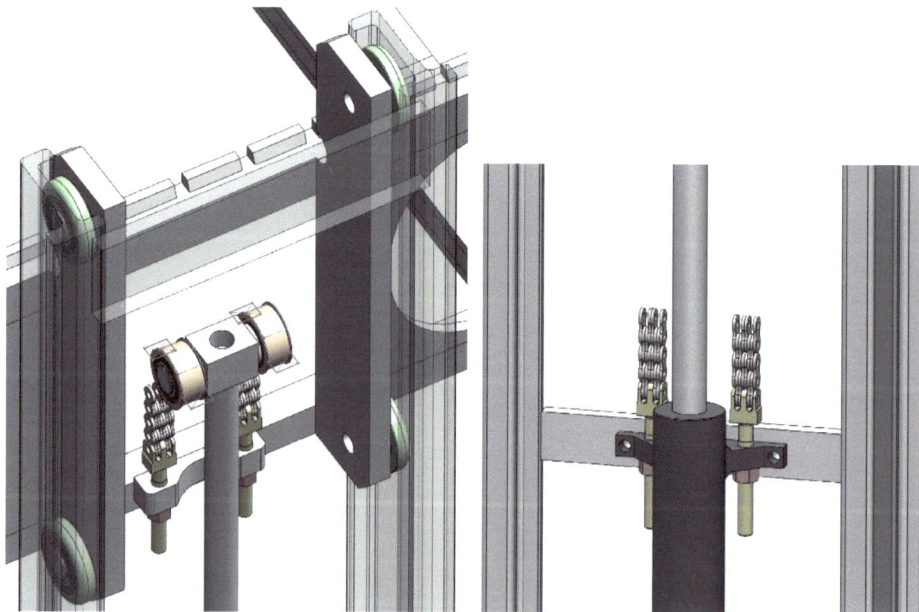

Free lift chain installation

A short section of each leaf chain end is shown in the model. The chain sheave model carries reference sketches of the chain outline for design purposes. The fixed ends of the free lift chains are tied to a block on the cylinder, where it is stabilized against the inner rail weldment.

Similarly, the "main lift" chains are run over sheaves hung on the middle rail weldment.

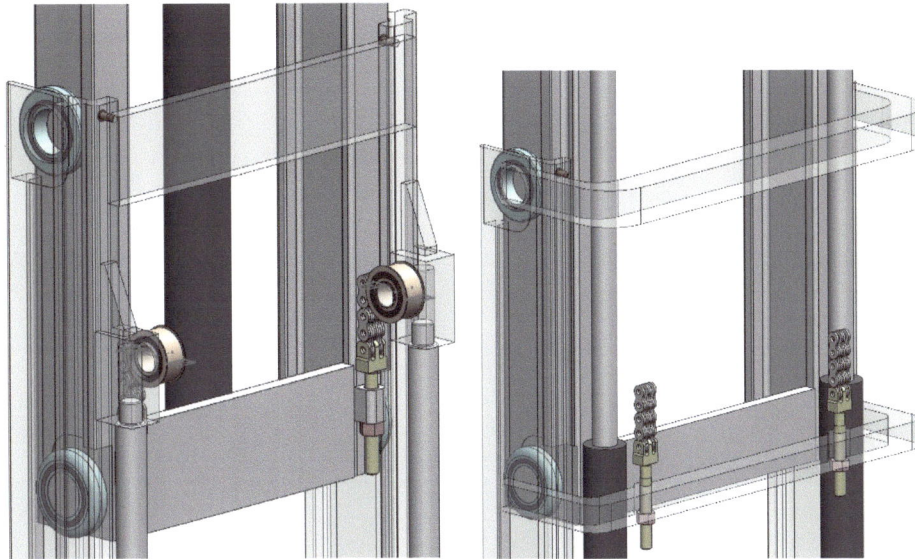

Main lift chain installation

These chains run from bosses near the bottom of the inner rail weldment to a crossmember on the outer rail weldment. In this arrangement the inner weldment lifts at twice the speed of main lift cylinder extension.

Mast Mounting

The mast is to be mounted on a lift truck so that it can pivot fore and aft.

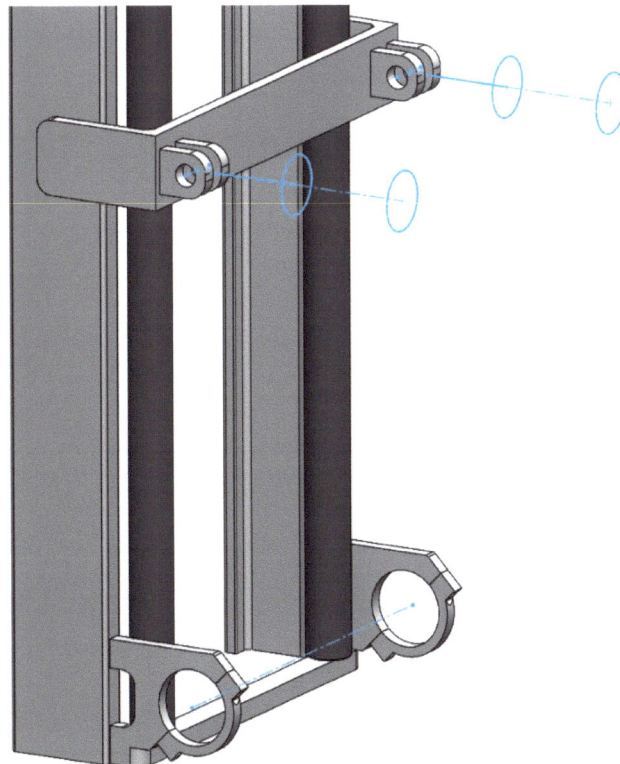

Mast-truck mounting with tilt cylinders sketched

The mast is supported at the bottom on a large axle, which may the truck's own true front wheel axle. A pair of tilt cylinders are pinned to double clevises higher up on the outer rail weldment. These cylinders control tilt of the entire mast.

New Assembly for Simulation

For these studies a new assembly is made with the mast model as the first part. Each of the main components of the simulation model is then created as a context part in that assembly.

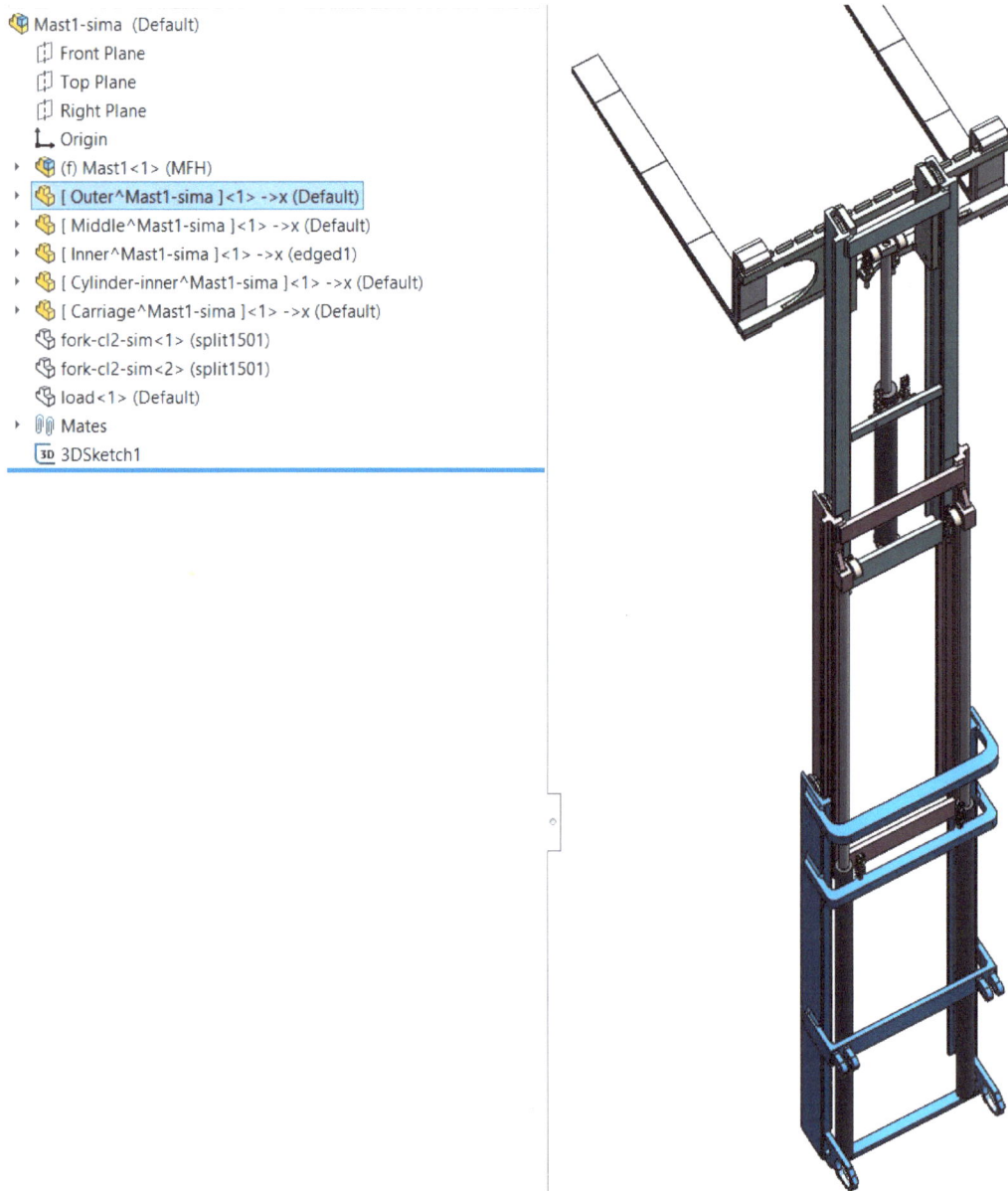

- Mast1-sima (Default)
  - Front Plane
  - Top Plane
  - Right Plane
  - Origin
  - (f) Mast1<1> (MFH)
  - [Outer^Mast1-sima]<1> ->x (Default)
  - [Middle^Mast1-sima]<1> ->x (Default)
  - [Inner^Mast1-sima]<1> ->x (edged1)
  - [Cylinder-inner^Mast1-sima]<1> ->x (Default)
  - [Carriage^Mast1-sima]<1> ->x (Default)
  - fork-cl2-sim<1> (split1501)
  - fork-cl2-sim<2> (split1501)
  - load<1> (Default)
  - Mates
  - 3DSketch1

New assembly with context parts for each weldment

The new parts are started with a "join" operation, components of each separate weldment merged into a single solid. The use of a new assembly is an optional step for file organization and to separate production design models from simulation models. The details are not covered here as this is outside the scope of this book.

Weldment Detailing

Since the rail sets and carriage are weldments, the weld beads and gaps between parts are added to the model as solid features.

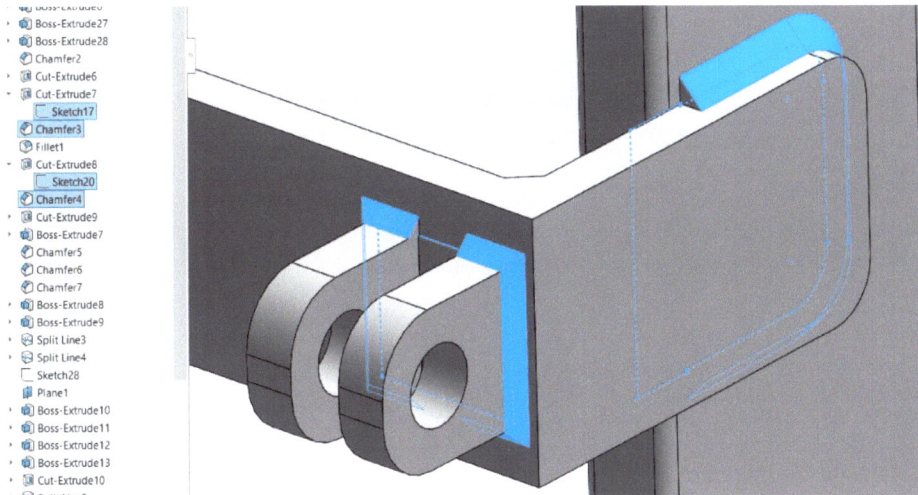

Weld details

The material properties of the weld filler material will be the same as the base material; a generic alloy steel is used here. Remember that only elastic modulus and Poisson's ratio are used in calculation of a linear static analysis. If the elastic properties of the filler and base material are substantially different (e.g. some cast irons) the welds can be made separate solid bodies and then bonded to the base material. Details of this methodology are covered in Solidworks Simulation for Real Weldments.

Rollers and Pads as Solid

Sliding pads and rollers are joined to the weldment models as parts of the solid. In a static analysis free rolling is identical to frictionless contact.

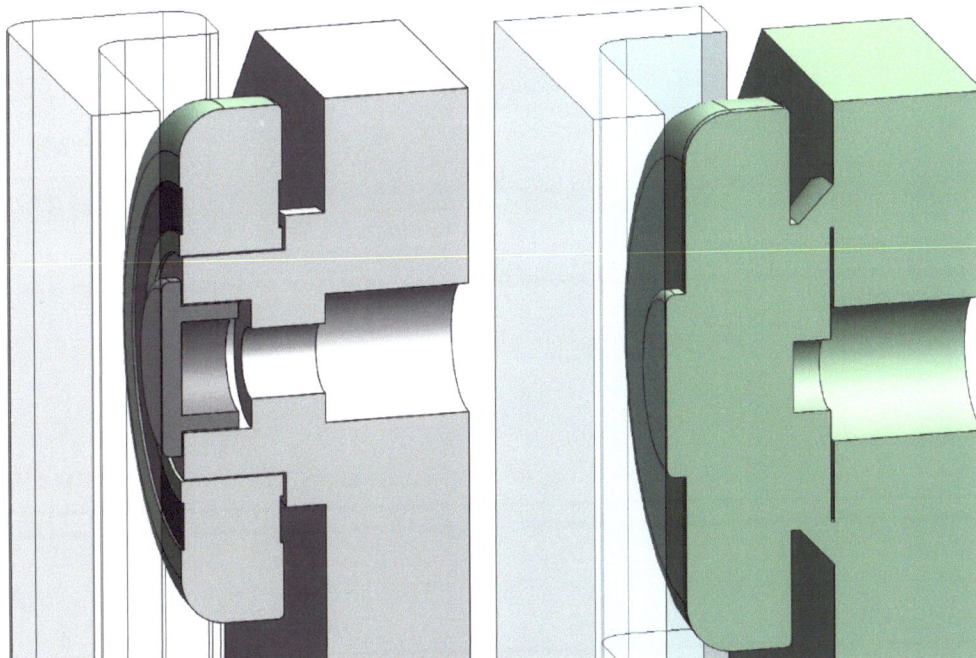

Carriage roller section views

A slight loss of fidelity is expected here if the sliding pad is made of something softer like a high lubricity plastic or impregnated bronze. It will have the same properties as the underlying weldment, in this case alloy steel. Here too the flat top of the pad could be made a separate body and bonded to the roller.

A few extrude operations are used to eliminate fine details from the roller and stub shaft and to eliminate small crevices which can give trouble in meshing. Also seen here is the stub shaft weld bead and modeled gap under that weld.

Initial Simulation Setup

The eventual objective is a simulation that accurately represents all loose solids, realistically connected and in contact.

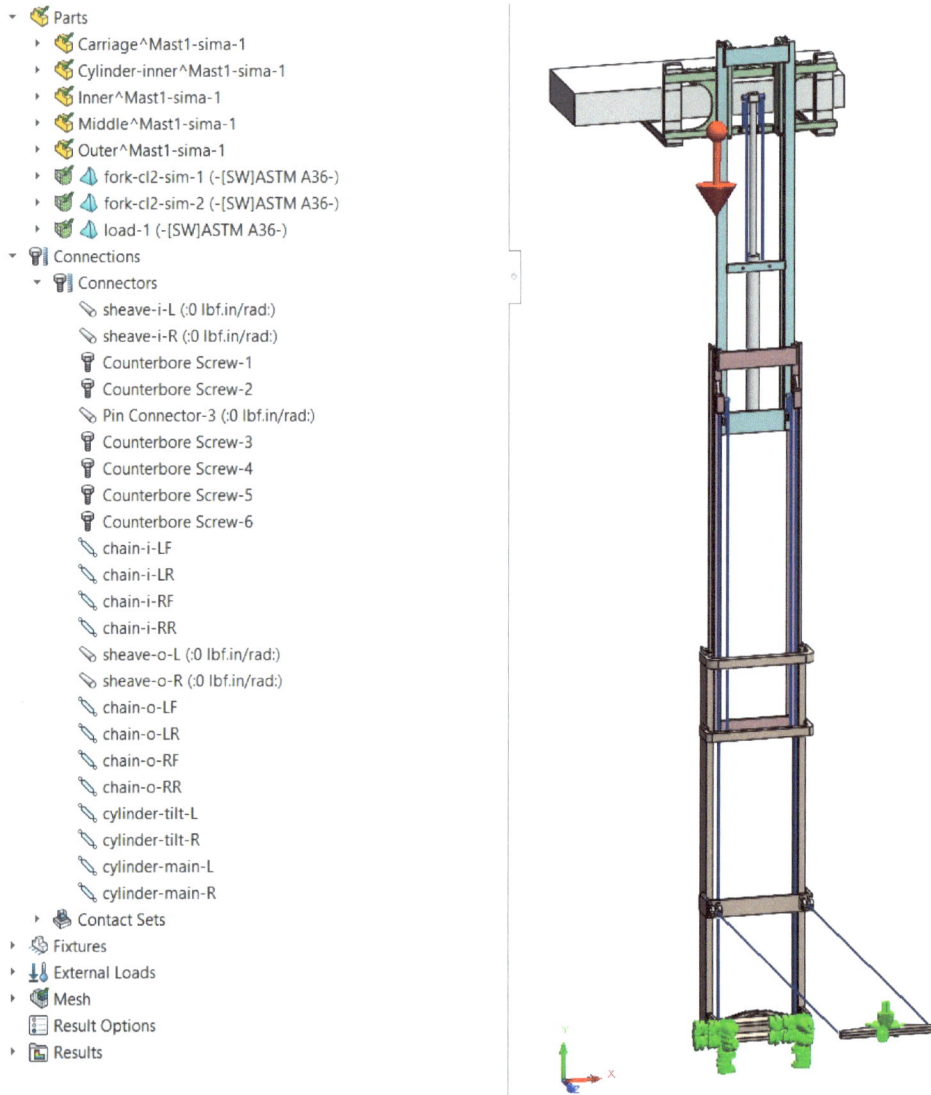

Complete simulation setup

Since there are so many contacts and connectors we do not expect to get everything working right on the first attempt. So, we start small. For a first pass only the carriage and inner rail weldment are included. Simple forks and a load block are included. Everything is in contact; no connectors are used.

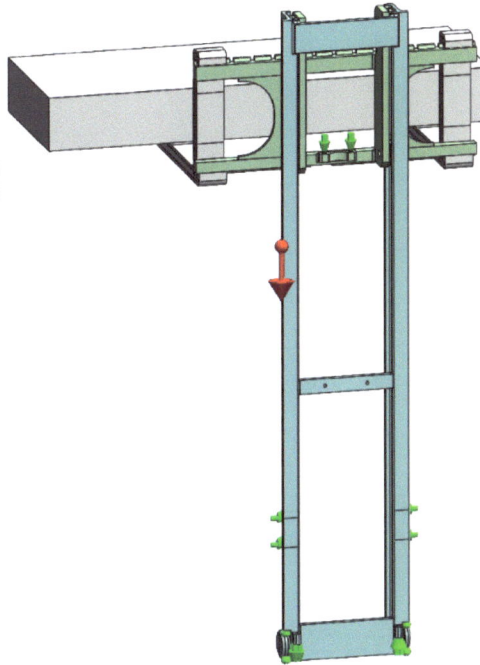

Initial study setup

Gravity Loading

For a lifting device gravity is the real loading. Modeling a load block of accurate mass allows us to place the block on the forks and use gravity as the only input. There is no need to compute new force vectors on the forks for different conditions. The single gravity vector can be varied in scale and direction instead.

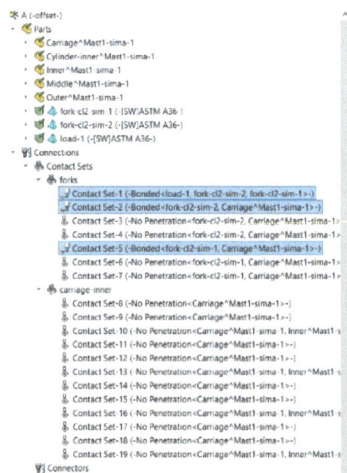

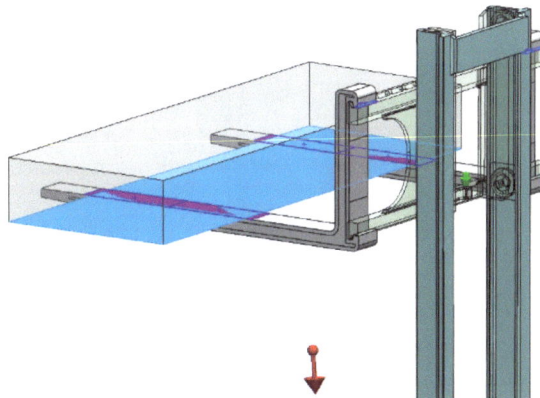

Bonded contacts on forks

The load block is positioned at rated distance from the rear of the forks (24 inches to center). For the initial studies it is offset to one side and the gravity vector at an angle biased to the same direction. This will give a more interesting response than a vertical loading and show us if our setup is working as expected.

The load block is "bonded" to the forks and the forks are bonded to the carriage at the bottoms of the upper hooks. These surfaces we are sure will always be touching and we have found having a few bonded contacts greatly speeds up a study over full no penetration contact. In the case of forks, if they were loaded with artificial forces they would be free to flex and sway independently, whereas a solid load will tie them together to at least some extent.

For the finest possible detail in response a separate sub-study can be run with full no penetration contacts on the forks, but our goal here is to get response of the entire mast.

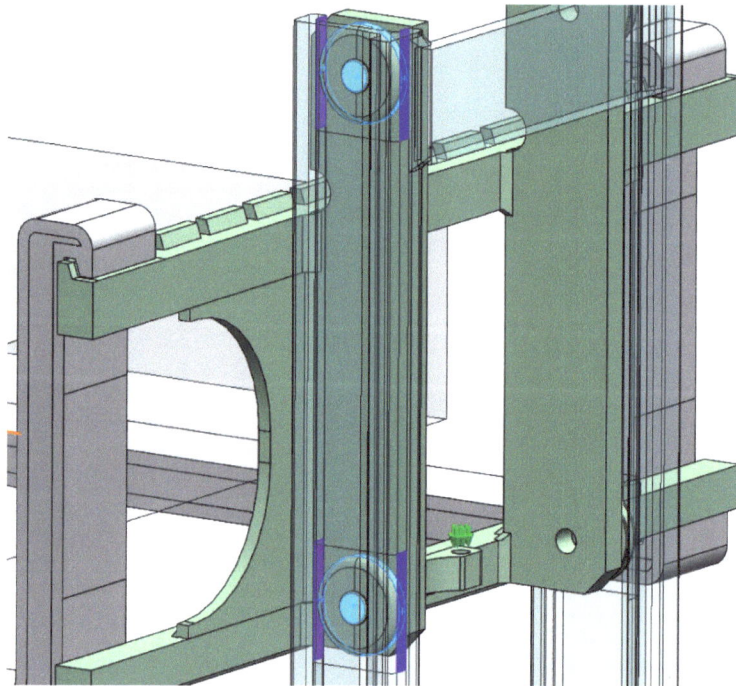
Roller, slider contacts from carriage to inner rails

Contact sets are defined between the carriage rollers and sliders and the inner rail weldment. Single surfaces are picked for each side of each contact set. The inner rail weldment surfaces were split ahead of time to reduce the area considered for each contact.

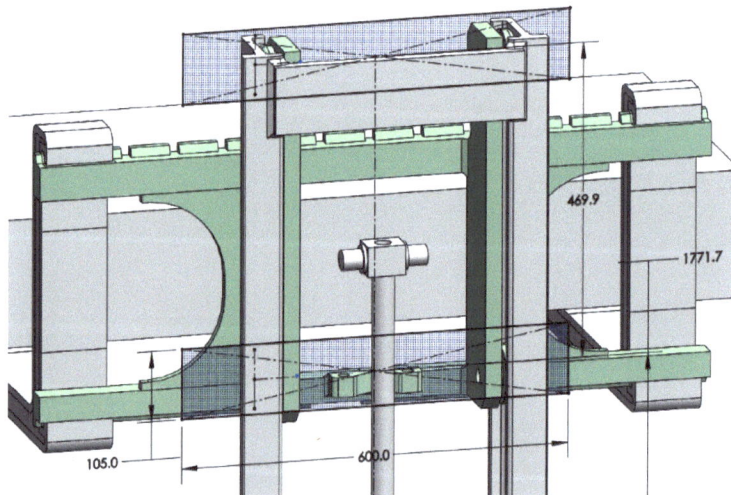
Sketch for split of inner rail surfaces

The split sketch is made with reference to the actual locations of the carriage rollers, but the sketch lines are ultimately located by free dimensions. In this way the locations of the splits can be changed or configured easily without unpredictable model updates forcing the inner weldment model to appear out-of-date as one moves between simulation studies.

For fixtures of this initial study all the lower roller contacts on the inner rail weldment are constrained in planar directions. The chain anchor locations at the bottom are held vertically.

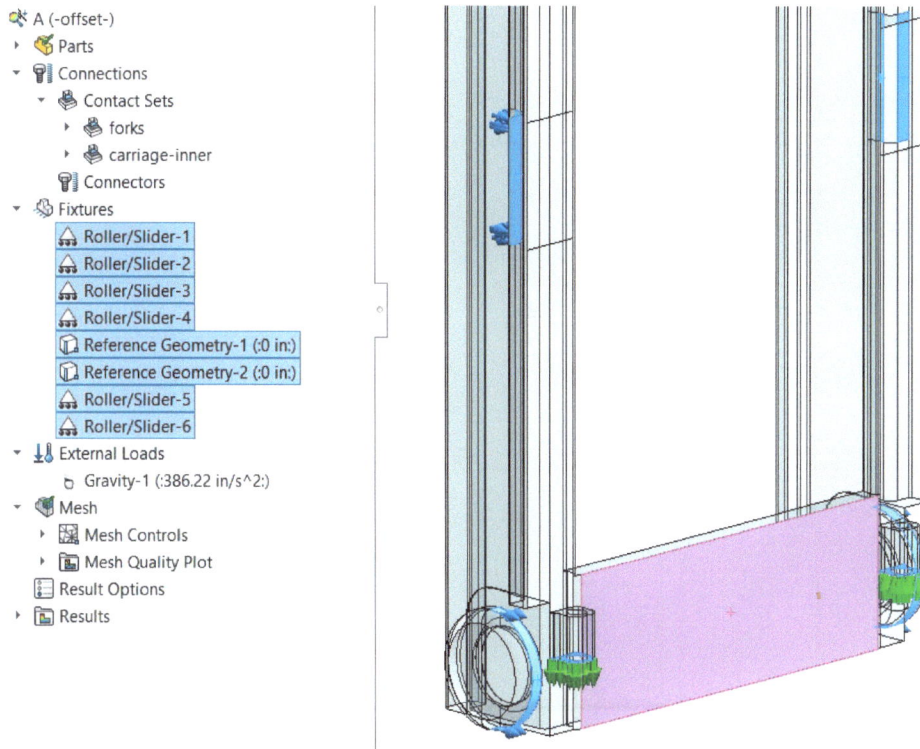

Fixtures on inner weldment

We have nothing yet in this setup to hold up the carriage. A constraint will be added at the chain anchors which normally support the carriage.

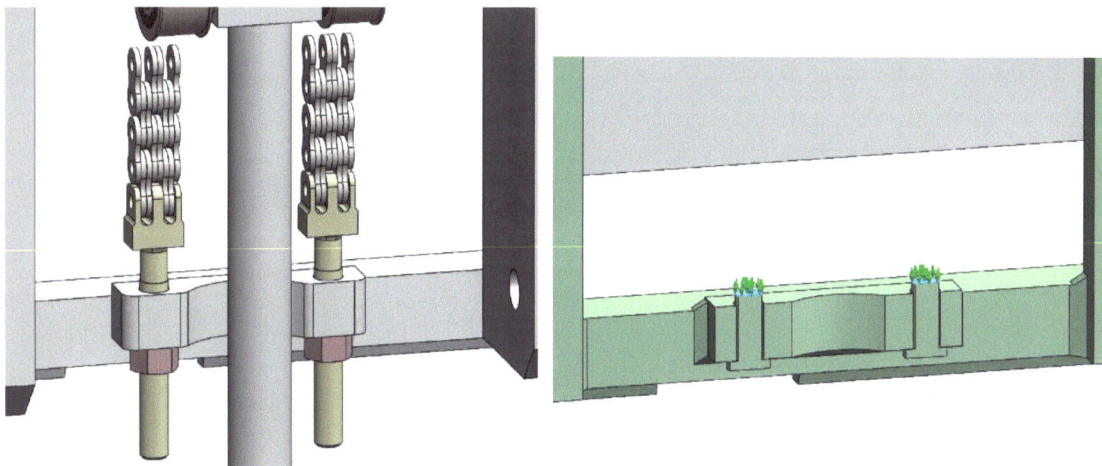

Chain anchor design model and simulation model section

A section view through the simulation model shows how the chain anchors will be handled. A simple stem and cap model of each anchor is modeled in place, with tight clearance to the through hole and touching the surface where the nut would bear. The stems are put in no penetration contact with the hole and bottom surface. For this study the tops of the anchors are fixed in place vertically.

Stress results appear generally as expected.

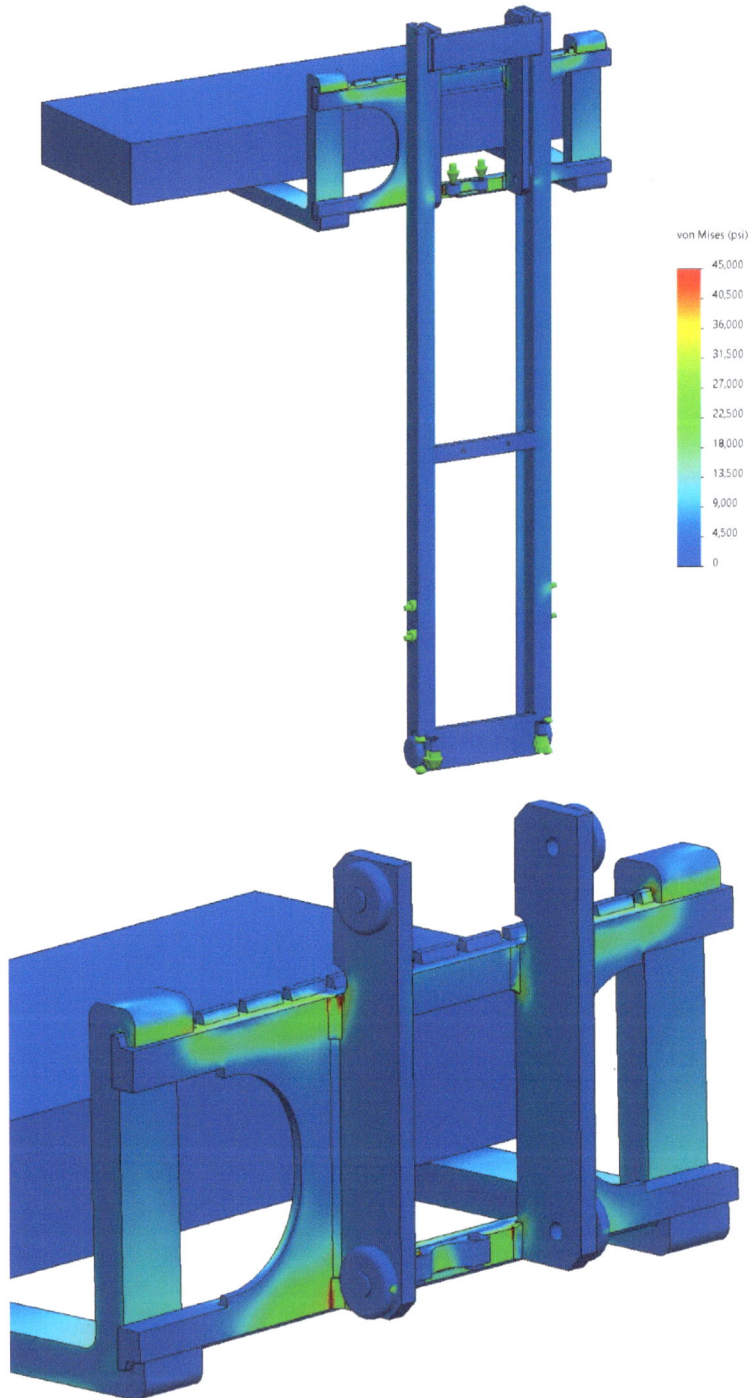

Model name: Mast1-sima
Study name: A(-offset-)
Plot type: Static nodal stress Stress1

von Mises (psi)

45,000
40,500
36,000
31,500
27,000
22,500
18,000
13,500
9,000
4,500
0

We focus mostly on the carriage at this point, as the setup on the rail weldment is temporary. Stress contours and peaks show a strong pattern from shear and bending forces on the fork bars. We could wait for the initial stress results to guide mesh refinement. In this case we used past experience to guide refinement on the fork bars, curved gusset plates, and on some of the weld beads.

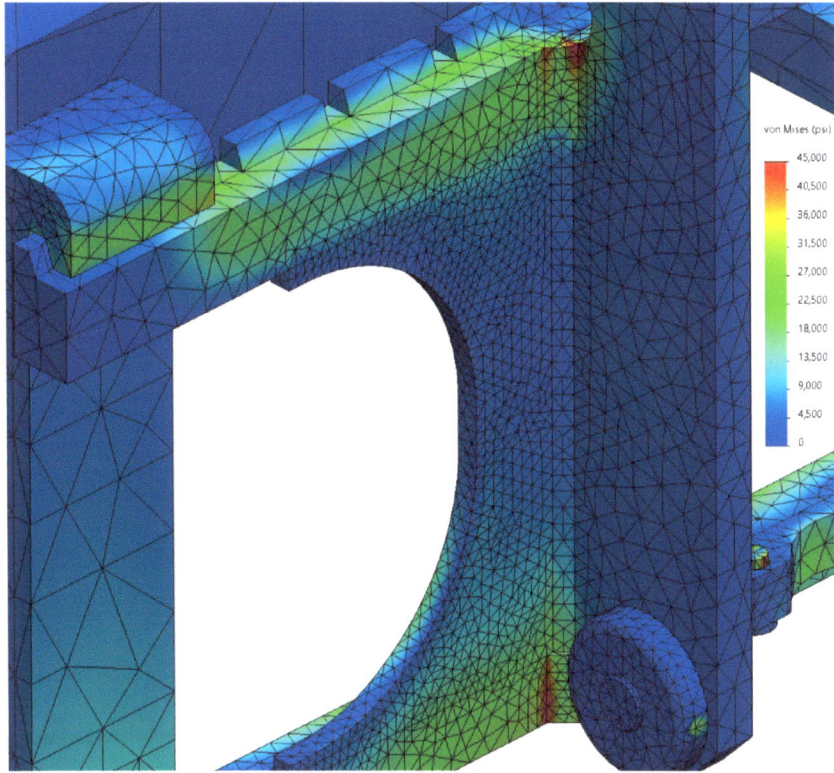

Results on mesh plot

We can check the forces coming in from the fork.

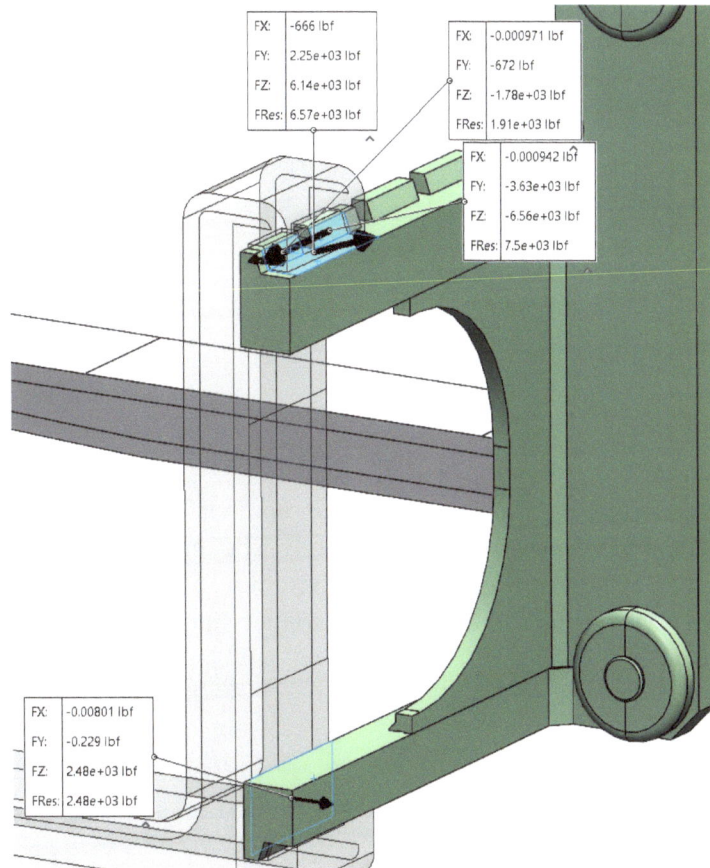

FX:	-666 lbf
FY:	2.25e+03 lbf
FZ:	6.14e+03 lbf
FRes:	6.57e+03 lbf

FX:	-0.000971 lbf
FY:	-672 lbf
FZ:	-1.78e+03 lbf
FRes:	1.91e+03 lbf

FX:	-0.000942 lbf
FY:	-3.63e+03 lbf
FZ:	-6.56e+03 lbf
FRes:	7.5e+03 lbf

FX:	-0.00801 lbf
FY:	-0.229 lbf
FZ:	2.48e+03 lbf
FRes:	2.48e+03 lbf

Forces on fork bar left ends

A "Free body force" check is used ("Contact/Friction force" does not work well with the bonded contact). The sum of vertical forces is a little over 2000 pounds downward, as expected. The fork rails are in bending from a 2480 pound rearward force at the bottom and an expected matching forward pull at the top. The bonded contact confuses the force results at the top, and even the most general "Free body force" inquiry we do not get expected values. We believe the forces are balanced internally; this is probably only a reporting issue in Solidworks.

Second Study Setup

Satisfied that the initial simulation setup elements are all working well, we delete the temporary fixtures and add in the free lift cylinder and the middle rail weldment. Typically we will make a new study but could also just edit the first one.

The new middle weldment is fixtured similarly to how the inner was done. Faces of the outer rollers and contacting faces on the middle weldment rollers are fixed in the contacting plane. The middle weldment has blocks into which the main lift cylinders would fit to raise the mast. The bottom surfaces of these blocks are constrained to stay at constant elevation.

The inner weldment needs to be held up, so its chain anchor blocks (near the bottom) are fixed in the same way as the carriage was previously.

The free lift cylinder is added as a separate body in contact. While we have spent a lot of time looking at way to represent hydraulic cylinders in detail, for this full mast study it is treated as one solid body. The rod, piston, and shell are merged into a single solid, simplified to remove internal detail. The shell portion is kept hollow to get the correct approximate mass.

Also joined to the free lift cylinder body is the crosshead. This small weldment or machined part carries the two free lift chain sheaves. Over these sheaves run the chains which support the carriage in the inner weldment.

We need to represent these chain sheaves and chains in a way which gives is the correct forces in the chains so that the weldments are loaded accurately.

Chain-over-sheave in Simulation

Running a chain over a sheave sets up a special kinematic relationship. At any instant we can consider each length of chain, off either side of the sheave, separately. The tension force in each is always equal to the other (a chain can also go slack, but we will let this go for now). The total length of the two chain segments is constant. This relationship can be represented by a simple mechanism.

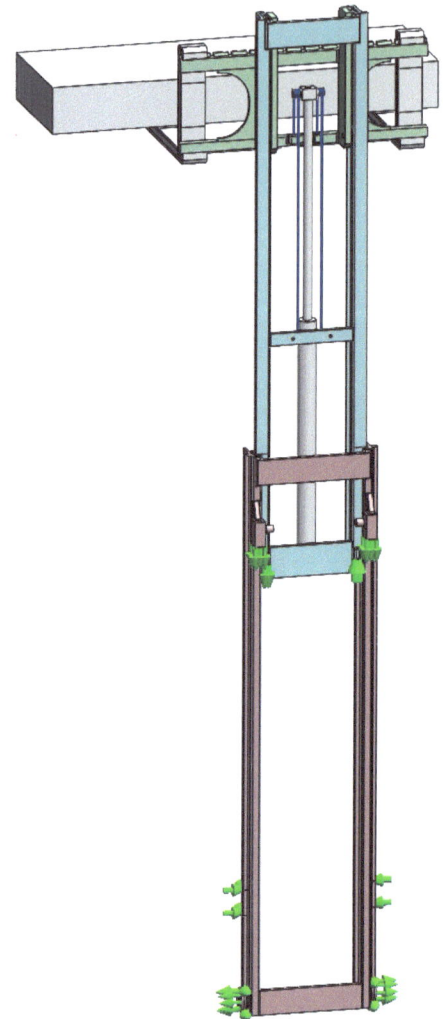

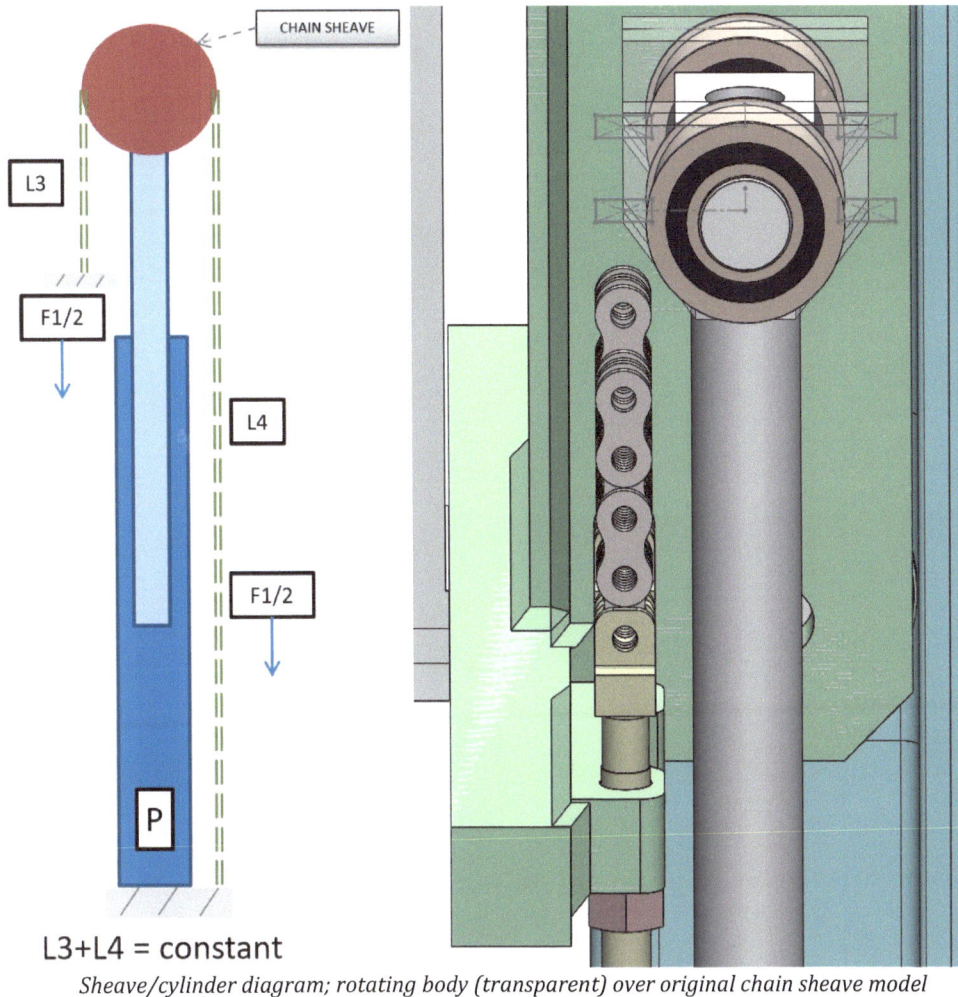

CHAIN SHEAVE

L3

F1/2

L4

F1/2

P

L3+L4 = constant

Sheave/cylinder diagram; rotating body (transparent) over original chain sheave model

Over a full lift or lower dozens of links of chains will pass over the each sheave. For a static simulation practically no length of chain will pass over the sheave. So we will replace the sheave with a simple rotating body with a vertex on either side at what would have been the tangent point of the chain centerline.

The new solids are positioned on either chain sheave shaft and articulated with a pin connector, with "retaining ring". Simple link connectors are then run between these rotating bodies and our simplified chain anchors.

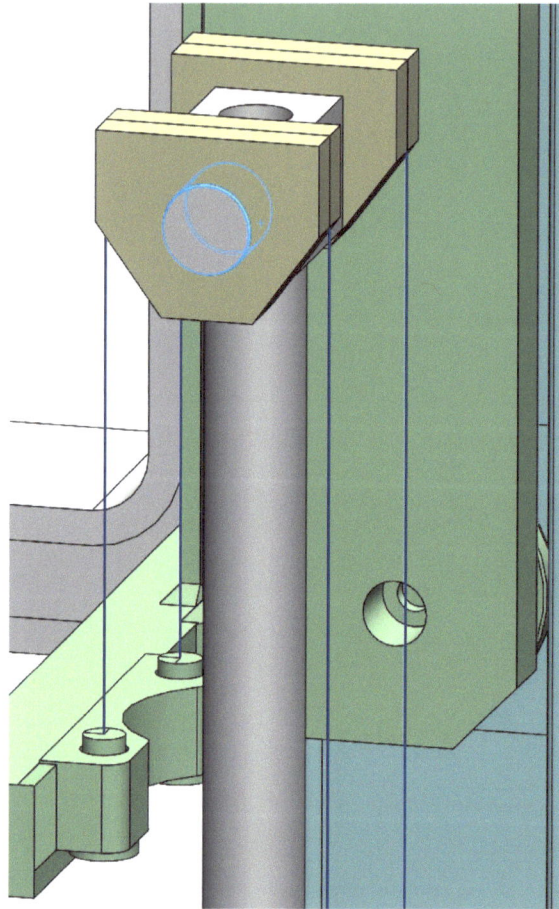

Simulated chain sheave and link connectors

The other ends of the longer link connectors are attached to anchors in a block on the cylinder shell.

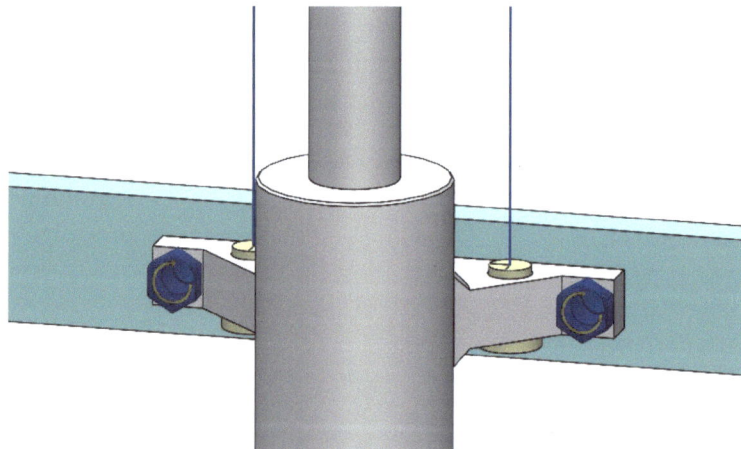

Cylinder shell brace

This block is bolted to a crossmember of the inner weldment. A contact condition is defined between the block and the crossmember and bolt connectors are set up in the two holes.

Here again we slip a bonded contact into the setup. We assume that the bottom of the free lift cylinder will always be pushing down on its support plate.

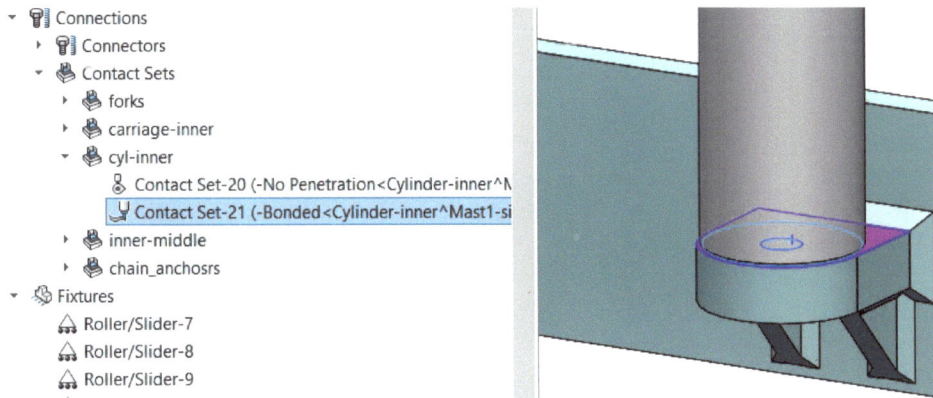

Bonded contact under cylinder

We expect judicious use of bonded contacts to meaningfully speed up the run. Even with this the study now takes over an hour to run, instead of about eight minutes for the first study.

Result with middle weldment and free lift chains

Support of the carriage is now completely flexible. The two lift chains can move independently and the rollers each have three separate contacts. We expect results on the carriage to be about the same as the final full mast result.

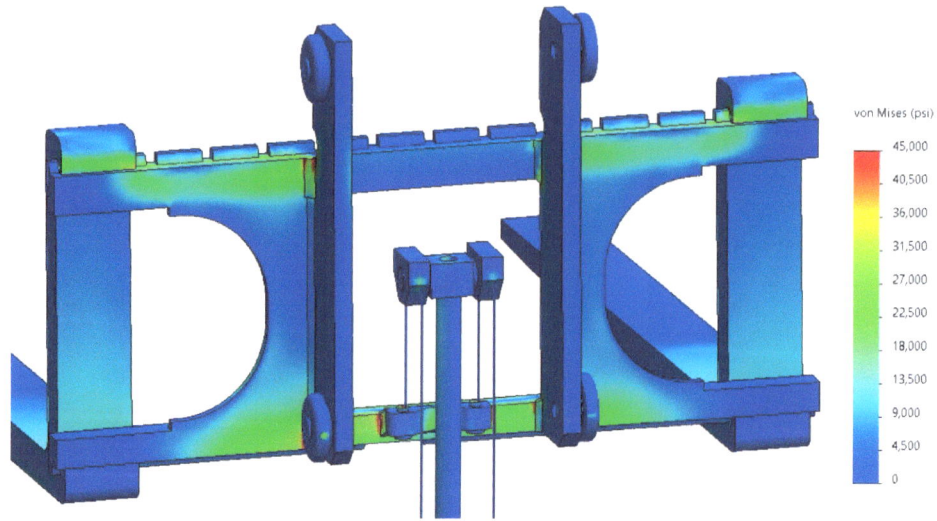

von Mises (psi)

45,000
40,500
36,000
31,500
27,000
22,500
18,000
13,500
9,000
4,500
0

Results on carriage with chains and rotating sheaves

We can check forces on the rollers and anchors to see what's going on.

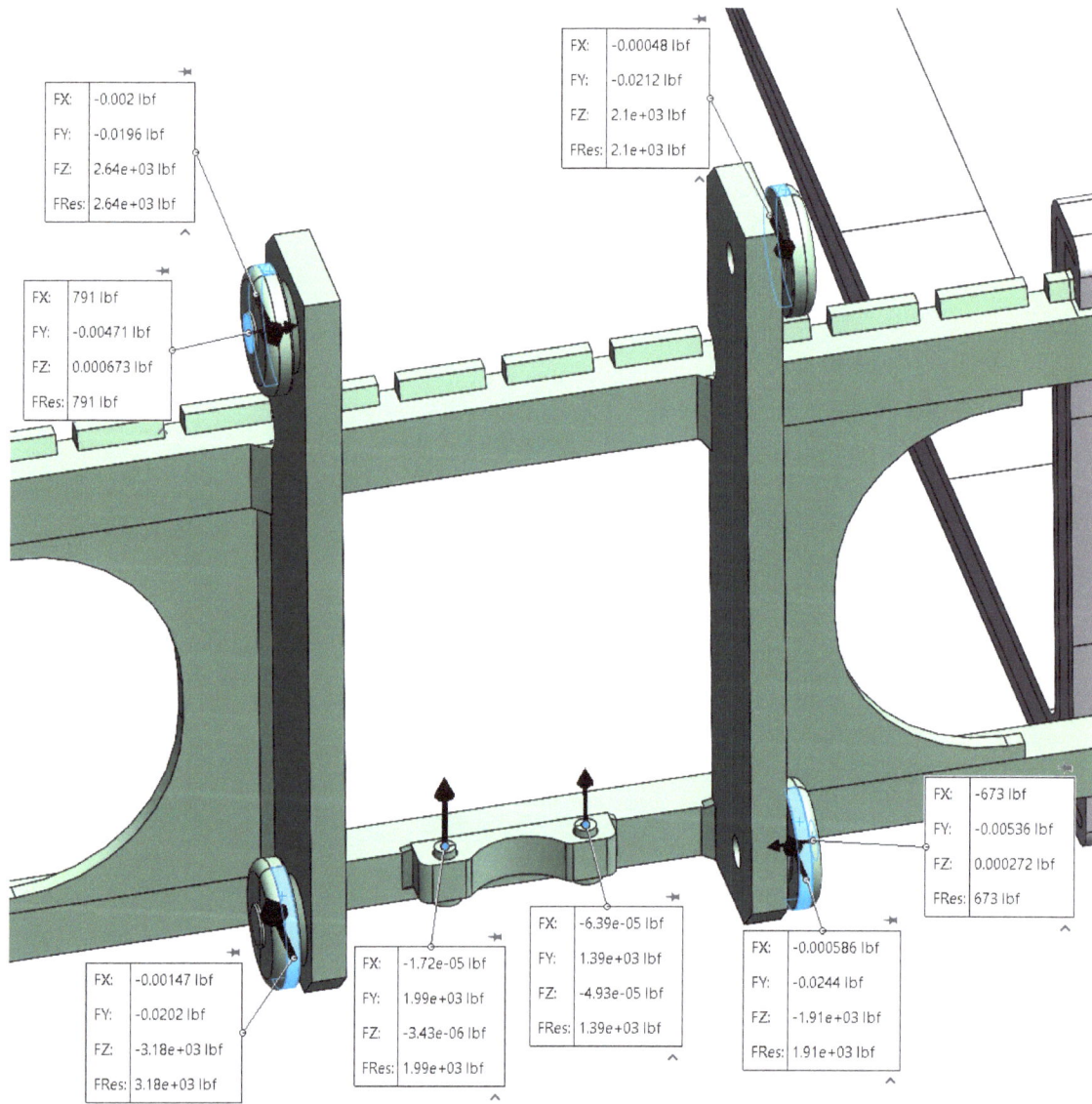

FX:	-0.00048 lbf
FY:	-0.0212 lbf
FZ:	2.1e+03 lbf
FRes:	2.1e+03 lbf

FX:	-0.002 lbf
FY:	-0.0196 lbf
FZ:	2.64e+03 lbf
FRes:	2.64e+03 lbf

FX:	791 lbf
FY:	-0.00471 lbf
FZ:	0.000673 lbf
FRes:	791 lbf

FX:	-673 lbf
FY:	-0.00536 lbf
FZ:	0.000272 lbf
FRes:	673 lbf

FX:	-0.00147 lbf
FY:	-0.0202 lbf
FZ:	-3.18e+03 lbf
FRes:	3.18e+03 lbf

FX:	-1.72e-05 lbf
FY:	1.99e+03 lbf
FZ:	-3.43e-06 lbf
FRes:	1.99e+03 lbf

FX:	-6.39e-05 lbf
FY:	1.39e+03 lbf
FZ:	-4.93e-05 lbf
FRes:	1.39e+03 lbf

FX:	-0.000586 lbf
FY:	-0.0244 lbf
FZ:	-1.91e+03 lbf
FRes:	1.91e+03 lbf

Free body forces on carriage rollers and chains

The offset and sideways loading gives an expected force distribution. A larger share of the overturning moment is taken up on the left side. The chain on that side carries more vertical load. Opposing lateral forces are seen in the upper left and lower right slider pads.

From these stress and force results we are satisfied that the study is running correctly so far.

Third Study Setup – Complete Mast

The outer rail weldment is added in. Contact conditions for the middle and outer rollers are established. These are identical to the contacts between the inner and middle weldments.

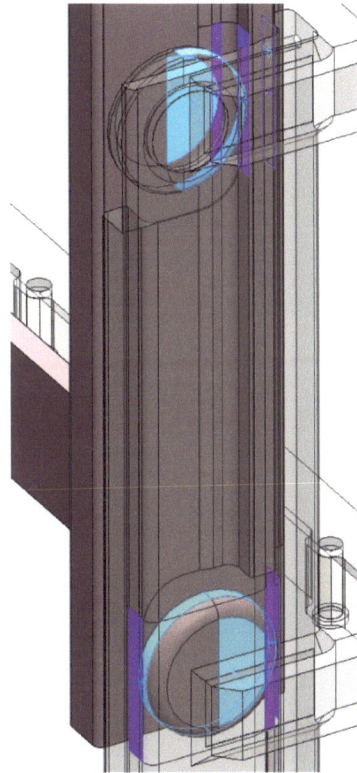

Contact surfaces, middle to outer

We are sure to catch the small slider behind the middle weldment, seen here in section view.

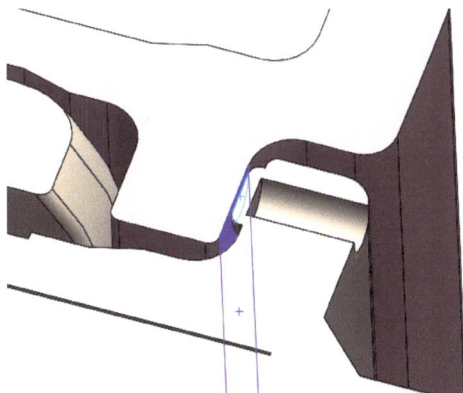

Rear slider contact surfaces

This contact adds theoretical stability, and we can't be sure when rear tilt or twist loading will bring a rail weldment back into contact with the rail behind it.

Cylinder Pairs

We have looked at some examples of fluid cylinders installed in machines. Often cylinders come in pairs, tied together to the same fluid feed. This sets up a unique kinematic relationship. Once both cylinders are extended to a desired length, and the control valve is closed, the total amount of fluid in the two cylinders is fixed.

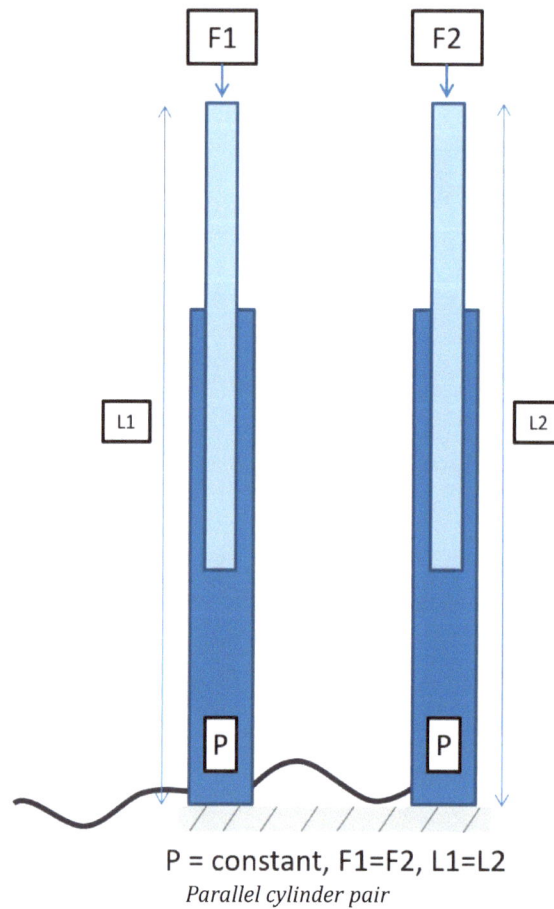

P = constant, F1=F2, L1=L2

Parallel cylinder pair

From that point if there is an asymmetric load put on the machine, the cylinders will react to the resulting movement. Some fluid may move from one cylinder to the other, so one cylinder gets shorter and the other longer. But once the fluid stops moving the pressure in both cylinders must be the same.

Mathematically the relationship established (for incompressible fluid) is: both lengths added together is a constant, and the force felt in one cylinder is equal to the force in the other.

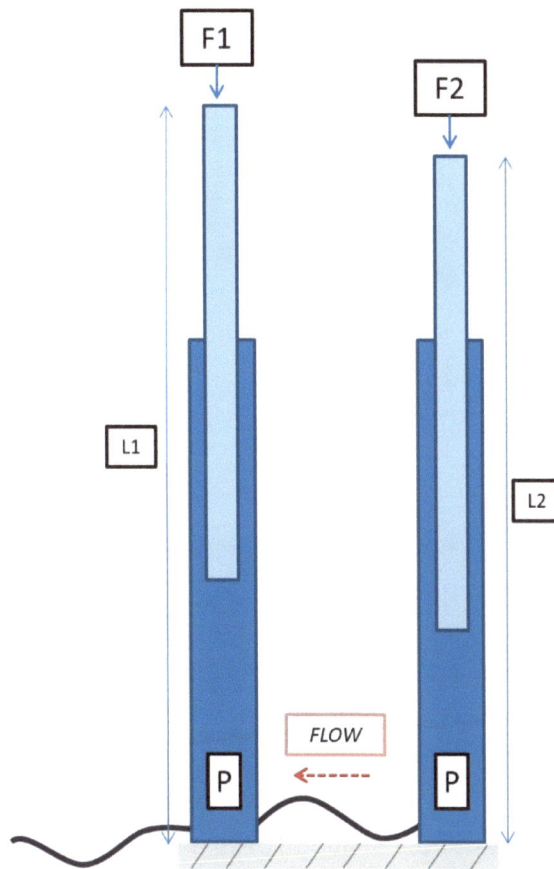

$(L1+L2)$ = constant, system
adjusts to maintain F1=F2

Parallel cylinder pair, rebalanced

A simple mechanical equivalent to this relationship is two very (infinitely) long parallel rods connected to a pivoting body. In simulation we can use link connectors for the rods.

Parallel cylinder simulation setup

Here two link connectors are run from clevis pins a long distance to a pivoting body. The pivoting body is on a "fixed hinge" fixture. In this simple linkage displacement to the right or left is passed freely from one link to the other, but the set can still resist total loads in that direction, net pulling or pushing, while the forces in the two links must be equal.

Tilt Cylinder Pair

Often a lift mast is mounted so that the entire mast can be tilted forward and rearward, usually by a pair of short hydraulic cylinders. The link/pivot setup can be applied directly to the tilt cylinders of this mast.

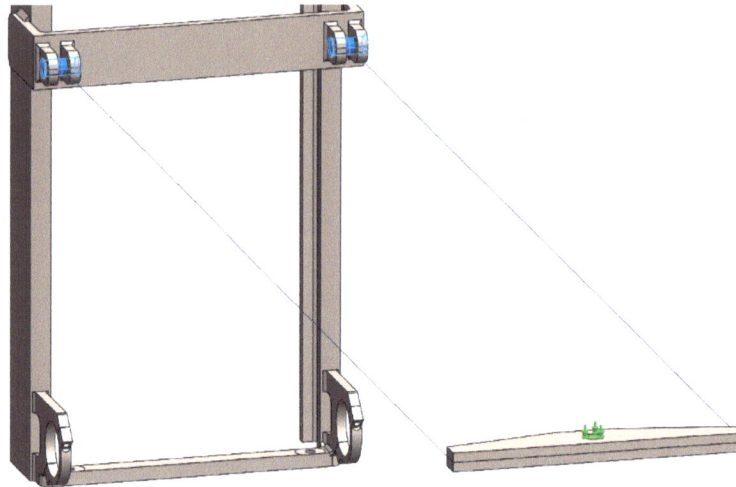

Tilt cylinder setup on outer mast model

In this case pins which would connect the tilt cylinders to clevises on the mast are included in the model. The pins are split to make a vertex at their center, to which the links are connected. Force will be concentrated at each connected pin vertex but spread through contact conditions into the clevises. As we saw on the small lift, these pins could be solid parts of the weldment with some loss of fidelity around the clevises.

Main Lift Pair

The rear pair of lift cylinders, also called the main lift, push between the middle rail weldment and the outer rail weldment. We need to allow them to move freely (in direct opposition) and still apply equal forces to top and bottom mating surfaces on both sides. So this is the same situation as with the tilt cylinders.

In this case the cylinders bear on parts of our assembly on both ends. If the bottom of the outer rail set is constrained, it can be a reasonable assumption that the bottoms of the main lift cylinders do not move relative to each other or to the outer weldment. In that case we can treat them the same way as the tilt cylinders.

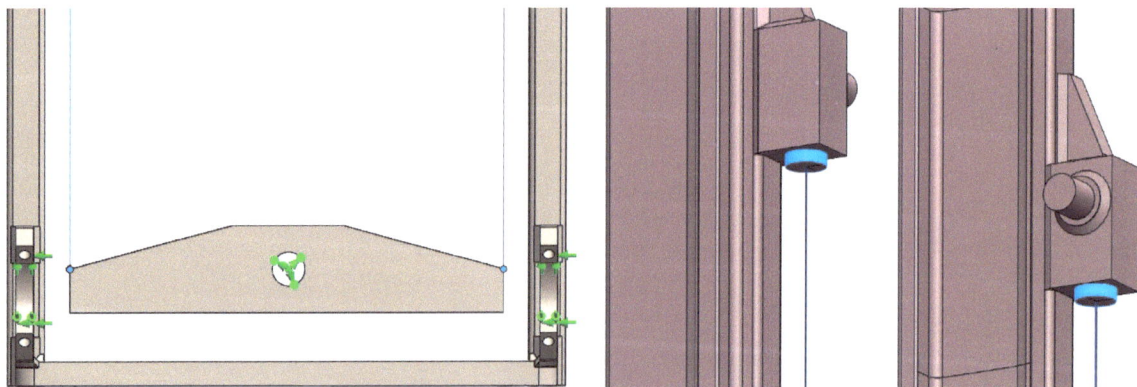

Main lift cylinder link setup, lower and upper

The bottoms of the links that stand in for our lift cylinders are joined to a pivoting body which is on a fixed hinge. The opposite ends of the links connect to the middle rail weldment at vertices created at the centers of the blocks where the cylinders would have connected. The expected force in each main lift cylinder (roughly double the load/fork/carriage/inner weight plus the middle weight) is applied manually to the outer weldment cylinder support bar. After the first full run, resultant forces in the link connectors can be checked and the manually applied force edited to match if needed.

With that we are ready to start the first full mast run.

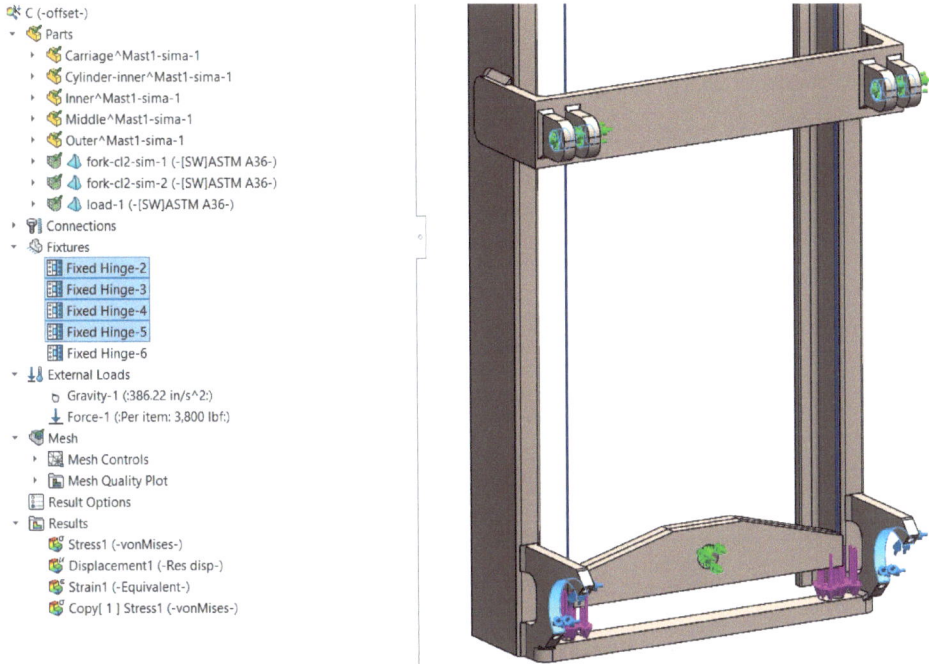

Setup at base of outer rail weldment

For now we leave off the tilt cylinder setup. The upper clevises are fixed on hinge conditions. The axle mounts are similarly fixed in place; those hinge fixtures work on half cylinders for further simplification.

We have been making mesh refinements along the way. Simulation model size is over 820,000 elements. This setup takes almost an hour and a half to run, compared to minutes for the first setup.

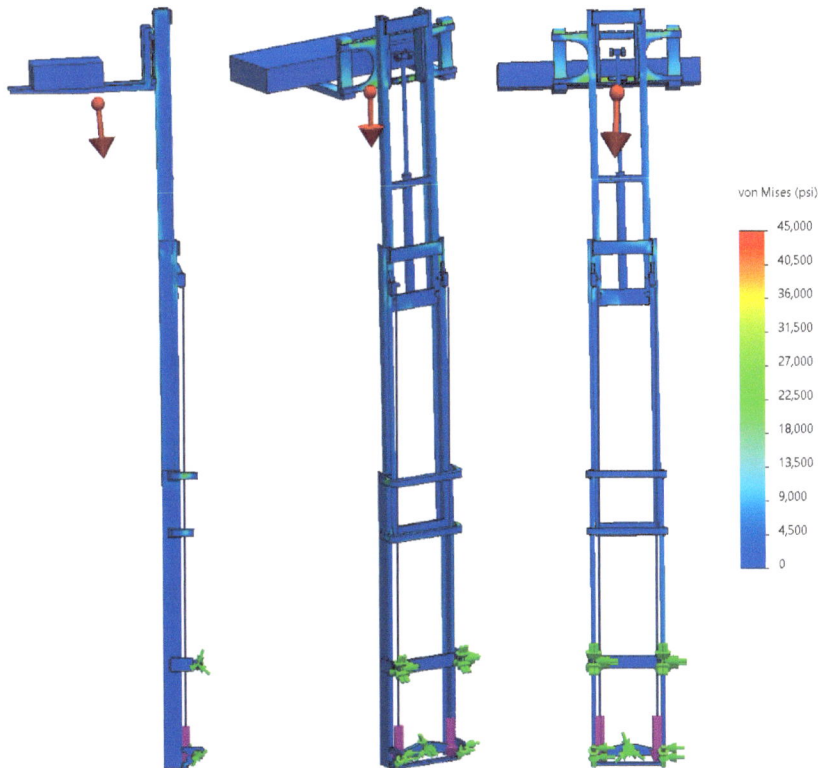

Results on first full mast model

With such a long open section supporting the load, at a significant moment arm, forward bending of the mast is noticeable even at a true view scale. Lateral lean is also easy to see. We look closer at the outer weldment, the component we just added to the study.

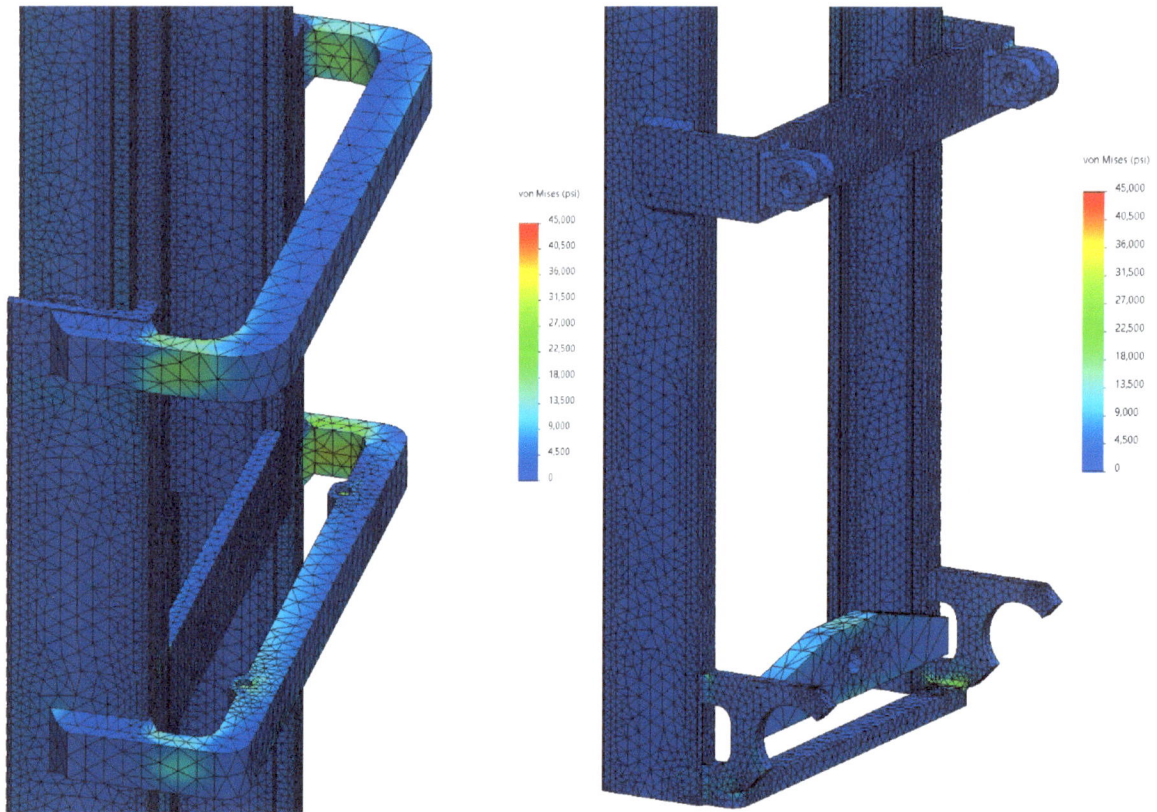

Results on outer weldment; upper, lower

The upper crossmembers are working hard, but reasonably stressed for low alloy steel [the stress values must be compared to actual material specifications of the crossmembers, the rails, and the weld filler as appropriate]. Stress levels in the fake pivoting body anchoring the main lift cylinder links are not important, so long as the body remains reasonably stiff. Stresses around the mounting parts are low, but we must remember that our boundary conditions were applied directly to surfaces on the weldment. This may artificially constrain local deformations and affect the stress pattern.

Finally we put the last pieces into our simulation model to get a mast that runs without any compromised boundary conditions forced onto the parts we care about. First, the flexible tilt cylinder arrangement is put back in as described earlier. Then a simple body is modeled to represent the truck axle.

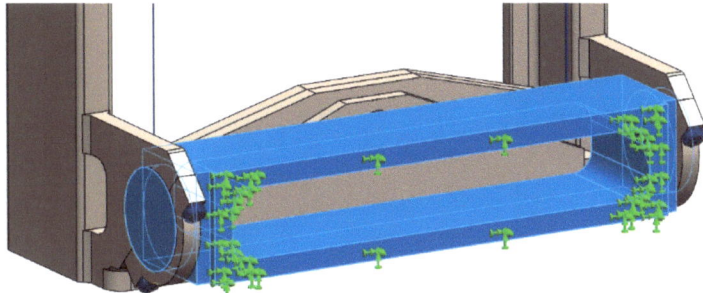

Truck axle body

The axle caps are put back into the model, in contact with the outer weldment and clamped by bolt connectors. The new axle body is in no-penetration contact with the now closed cylindrical axle mount faces and the inside faces of the mounting plates.

The main lift cylinders remain represented by link connectors joined to a pivoting body. But now we have introduced something complicated to literally bridge the gap between the floating pivot body and the cylinder support bar on which the cylinders are supposed to bear.

A new solid is added which spans the locations on which the cylinders would rest and which comes up far enough to cover the existing pivot body in a new double clevis.

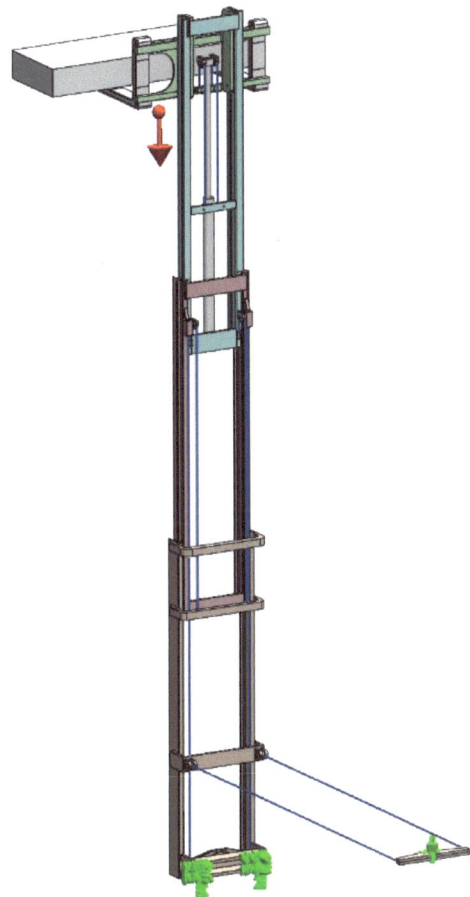

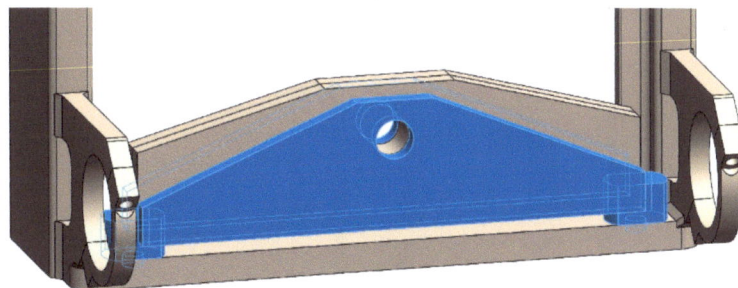

Main lift cylinder bridge

This bridge will take vertical force from the center of the pivot body and divide it equally between the locations where the real lift cylinders would rest. Such a body could stiffen the real parts considerably so it is located somewhat loosely with contact conditions into the holes through which the real cylinders would have met retaining screws.

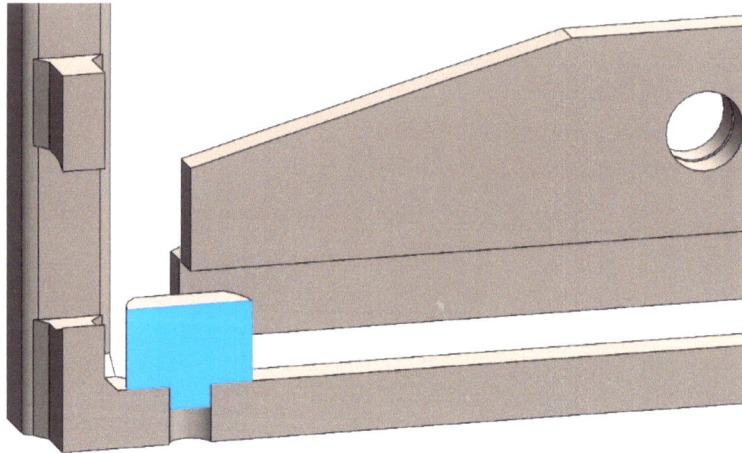

Section through bridge and main lift pivot

Now we do not need to manually apply force to the cylinder support plate. The only load input is gravity.

Overall response is similar to the previous run, which is not surprising.

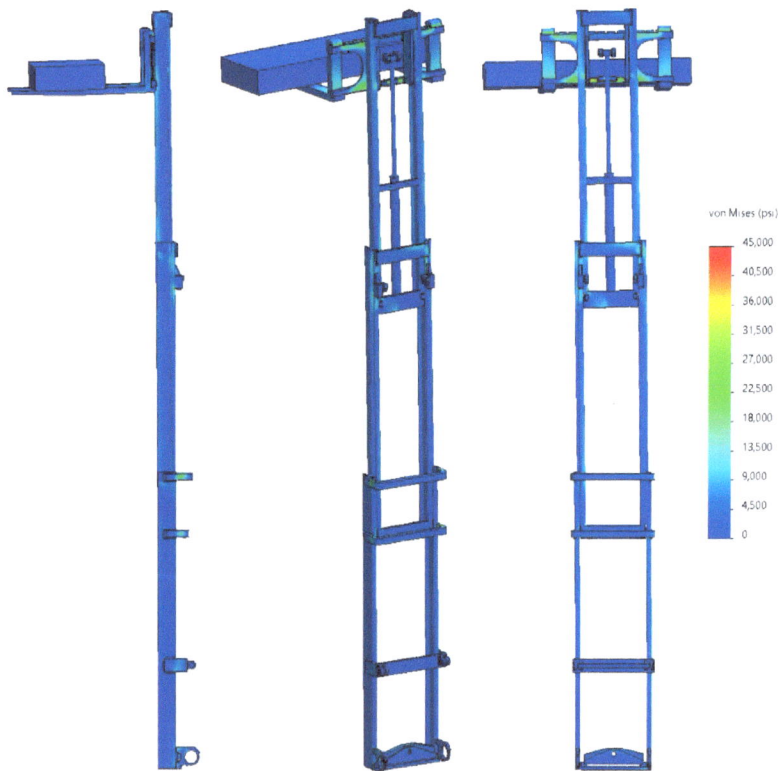

Result with flexible mounting

Differences become clearer in a top view. The outer weldment is now allowed to twist and move realistically on the axle; this motion is magnified through the length of the mast.

Solidworks Simulation for Real Machines

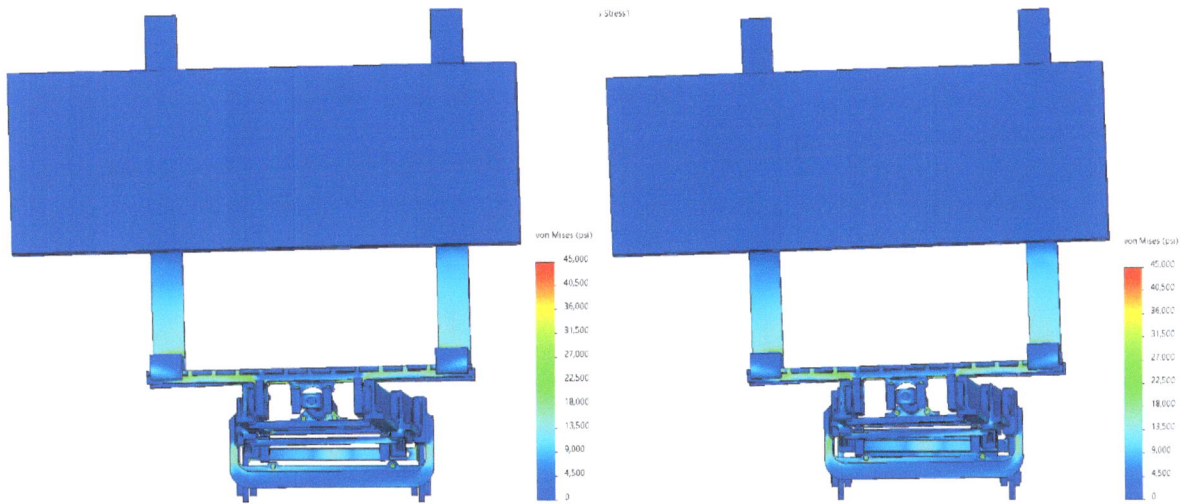

Result with close fixtures vs flexible mounting

A closer look at results near the mast mounting show that stresses did change in the weldment.

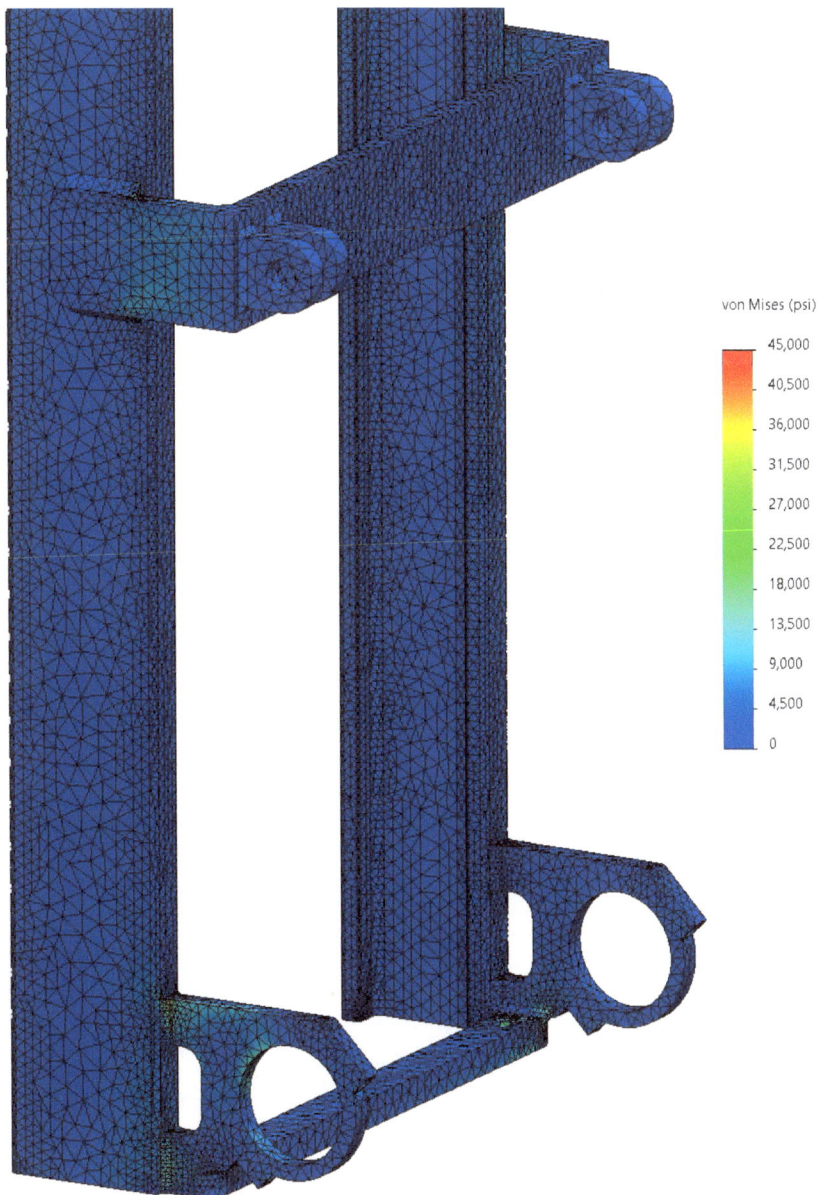

Result on lower outer weldment

Some mesh refinements were also made in the weld beads, so the results are not strictly comparable. But we are satisfied that this setup is the way we want to run this mast. With use of close-fitting solid "rollers", many discrete no-penetration contact conditions, a handful of bonded contacts, rotating virtual chain sheaves on pin connectors, link connectors for chains, joined links for parallel hydraulic cylinders, and a solid mounting axle we have a fully articulated model that moves realistically and gives stress results in which we have confidence.

The cost for this fidelity is that the study now runs with over 860,000 elements and takes well over an hour to run on a decent computer. We will look again to see what might be driving up the study size.

Outside Fillets in Curvature Based Mesh

We have been using the curvature-based mesher. It works well and quickly for the detailed weldment models we prefer. But its automatic refinement on fillets is a double-edged sword. Small fillets in the model can drive up mesh count much more than necessary.

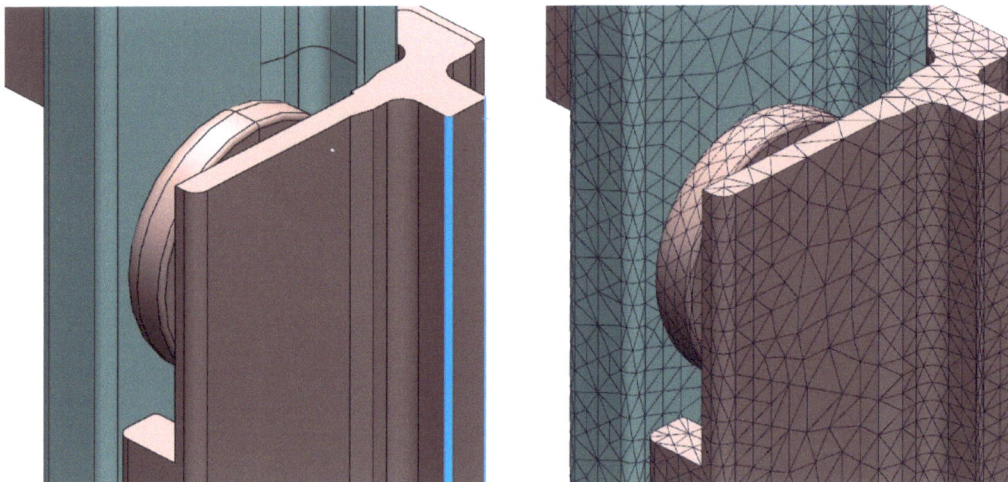

Small radii on rails and resultant mesh

The rail profiles in our model have some small outside fillets, 2mm radius in this case. These force the mesher to make small elements along the entire length of each rail, at each small radius. It is simple in this case to sharpen the edges with new extrude operations. The small edges are sharpened on all three rail sets.

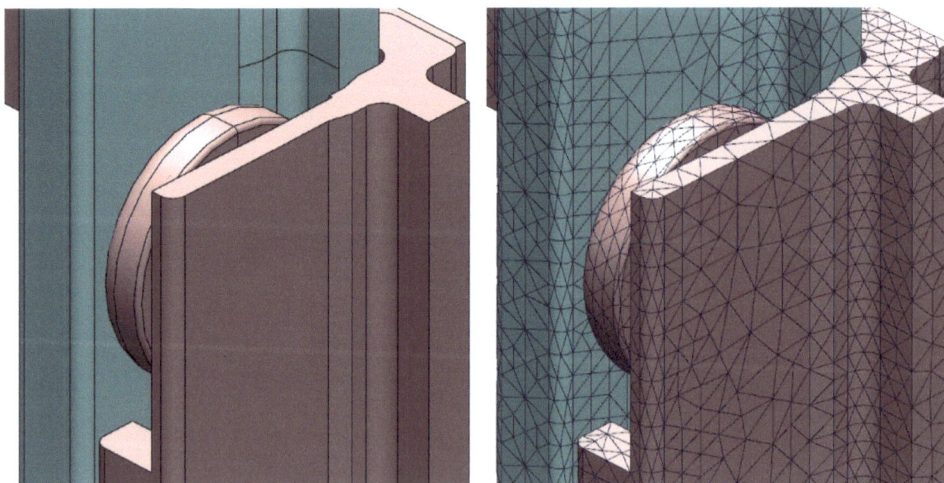

Model and mesh with small radii sharpened

The new total mesh comes to about 775,000 elements, down from 823,000. This is probably helpful in reducing run time without affecting results. To see what more can be accomplished the next larger outside radii on the rails are sharpened.

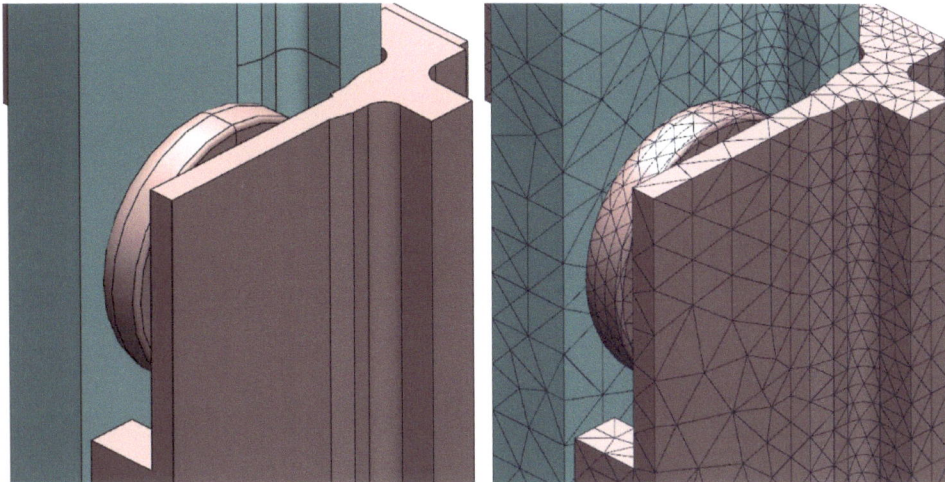

Model and mesh with larger radii sharpened

More progress is made on reducing the simulation problem size, as the new mesh is under 530,000 elements. But now we wonder if the results have been affected. The coarser mesh may be stiffer under load.

All three are run and selected displacements compared. The top center point on the back of the inner weldment is probed for forward and lateral displacement. Results are tabulated, scaled to the high mesh count run. [The first three studies in this project were labeled A, B, and C; studies for this meshing sub-project runs are labeled D, D2, and D3.]

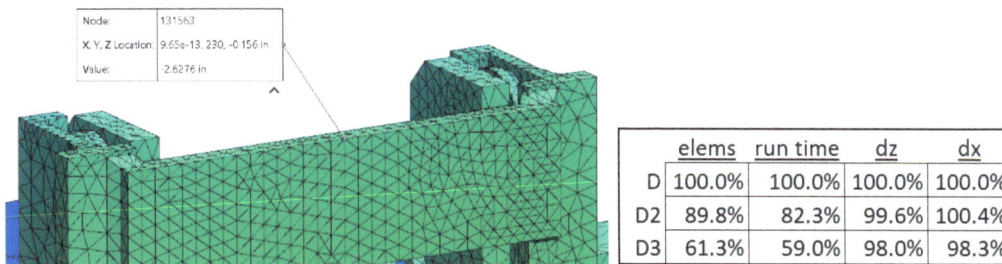

	elems	run time	dz	dx
D	100.0%	100.0%	100.0%	100.0%
D2	89.8%	82.3%	99.6%	100.4%
D3	61.3%	59.0%	98.0%	98.3%

Displacement probe and results summary

Reducing the mesh size by sharpening the radii has expected results on run time. The smaller mesh brings a result much faster. The coarser meshes are stiffer, as expected, but the difference is small. The fully sharpened shapes are less than two percent stiffer, in total over the whole mast. The effect probably has more to do with increase in the section modulus of each rail (we did add material at the extreme edges) than the changes in the mesh density. We will carry on with the sharpened model.

The extrude operations on the rails changed the split faces for contact. So those faces had to be re-split and the contact sets fixed. Once that chore is done we can duplicate the last meshing experiment study to make a new series, the actual runs we will use to evaluate the product.

Production Studies
We will look at the mast loaded by gravity in four conditions, labeled by number:

0. Unloaded vertical,
1. Loaded vertical,
2. Loaded forward lean (emulates fast stop or actual forward tilt),
3. Loaded rear/left lean (emulates turning and braking while backing).

The first study, unloaded, is mostly a check of our setup. In this condition structural deformation is expected to be minimal. Most displacement of the model should be from take-up of the clearances we left between components. If those clearances are realistic, and small, the result should be comparable to what is observed on a real build.

Loading by Gravity Vector Sketch

A 3D sketch at the assembly origin is used to define the gravity vectors. A dimensioned pyramid has lines running at angles up to the origin (it is most convenient if the lines are all ordered in the same direction).

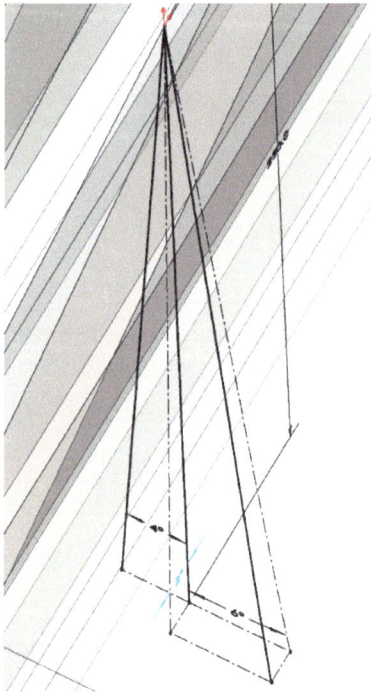

Sketch lines for gravity definition

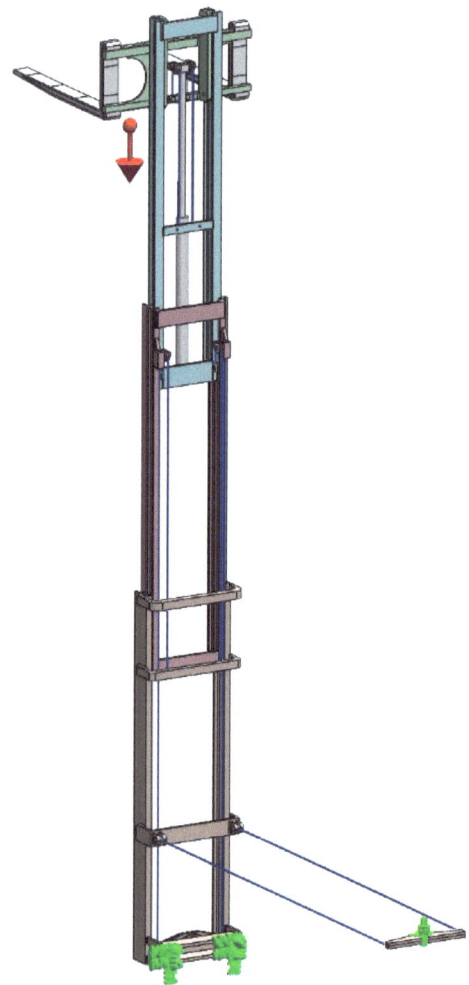

For these studies we defined vertical, four degrees forward, and a compound four degrees rearward with two degrees left as our dimensions for the gravity vector.

Studies 1 and 2 have the load block centered, mast vertical or tipped forward. Study 3 uses the configuration with the load offset to the left, for increased twist on the mast in the compound angle gravity loading. We can plot results from many angles to compare the response.

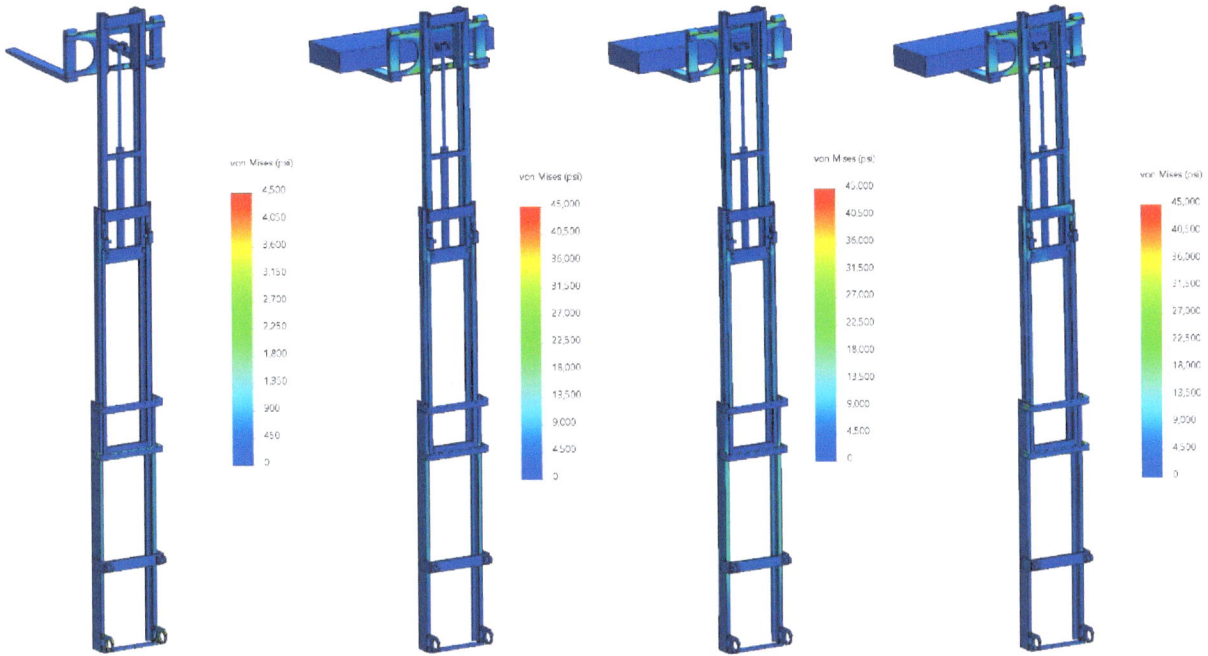

Stress plot (true displacement), rear quarter iso view

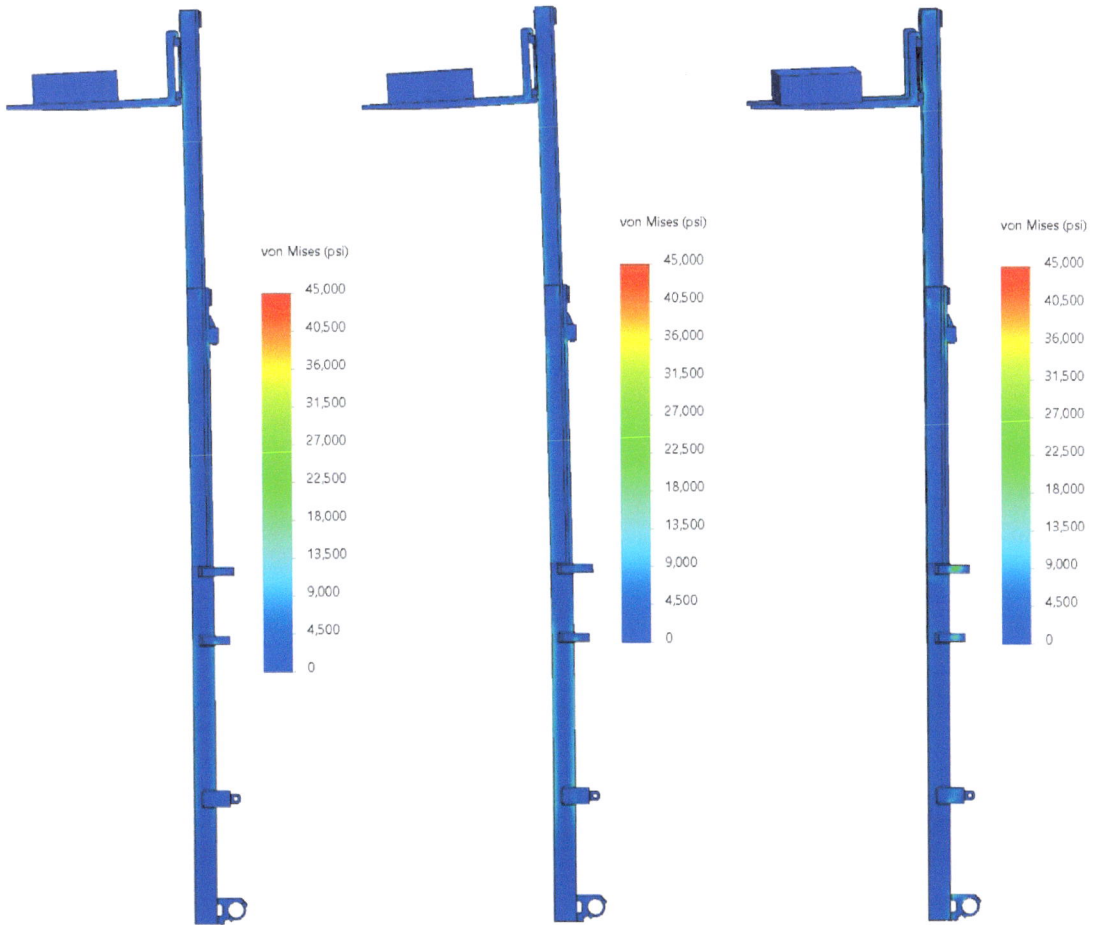

Side view

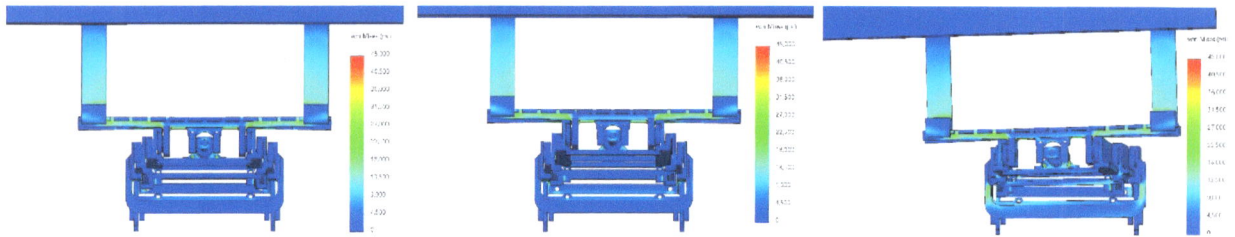

Top view

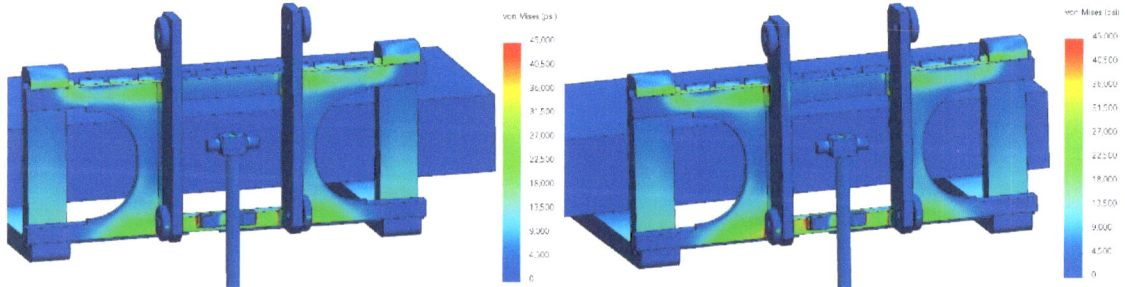

Result on fork carriage, centered vs offset load

Because we took care to retain detail in the model (including welds), to accurately represent relationships between moving parts, and to add mesh refinement where needed we can zoom in on any part of the model and expect usable local stress predictions.

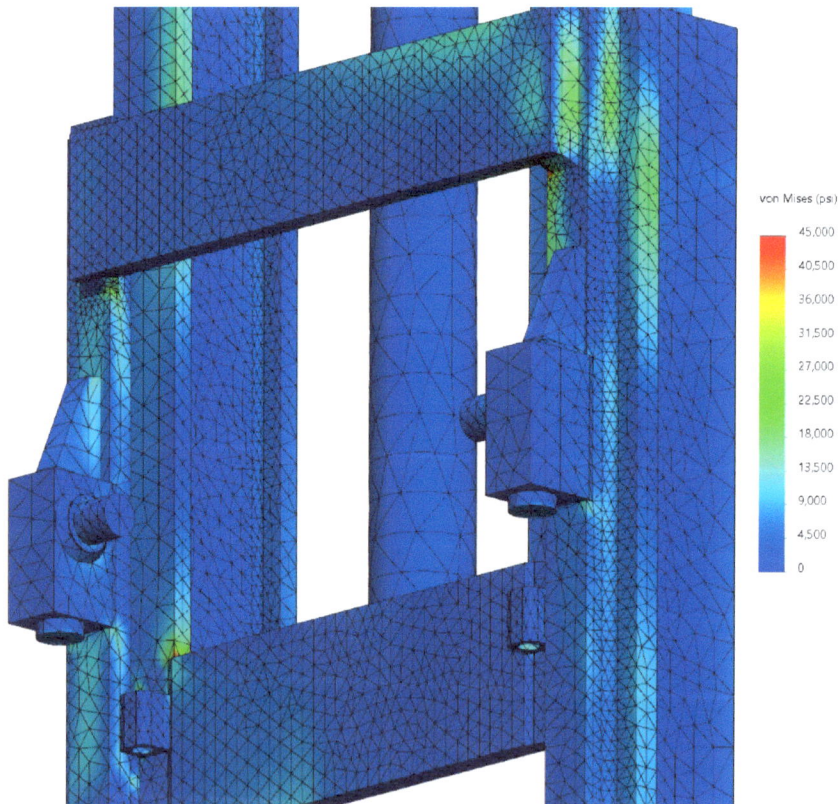

Result plot on mesh, detail view

The results are summarized with key values below. We recommend keeping a list like this for any set of studies, with additional columns for setup details, mesh statistics, and key results. It can greatly help identify unexpected behaviors.

	load	offset	fwd	left	dz	dx
0 vertical, unloaded	0	0	0	0	0.218	7E-04
1 vertical, loaded	3000	0	0	0	2.563	0.031
2 forward lean/stop	3000	0	4	0	3.953	0.107
3 rear/left turning/stop	3000	33%	-4	2	1.511	2.627

Summary of production runs

Deflection Results Caution

We can offer some cautious advice about using deflection results from an assembly model like this. Total deflection results are usually lower than observed on a real machine under load. Some of this is from normal and well documented 'stiffness' of finite element methods (a discussion of meshing and displacement results, with many experimental series, can be found on the Stone Lake Website under "Meshing Challenge"). Some of the low flexibility will come from our setup.

This model was built with small clearances between components. This represents an ideal build condition, and it is faster to solve in contact analysis. The mounting system, even with literal solids modeled, is less compliant than any real truck or test stand mounting. For this build it may be important that we represented chains and fluid-filled cylinders as rigid links.

Checking the Chains

A chain or rope or cable is an object which can take load practically only in tension. It can't be used to push or apply (much) lateral force. In the mast assembly we used rigid links for the chains. We need to check manually to see that they all stayed in tension. If a real chain goes slack, it will not support any pushing force.

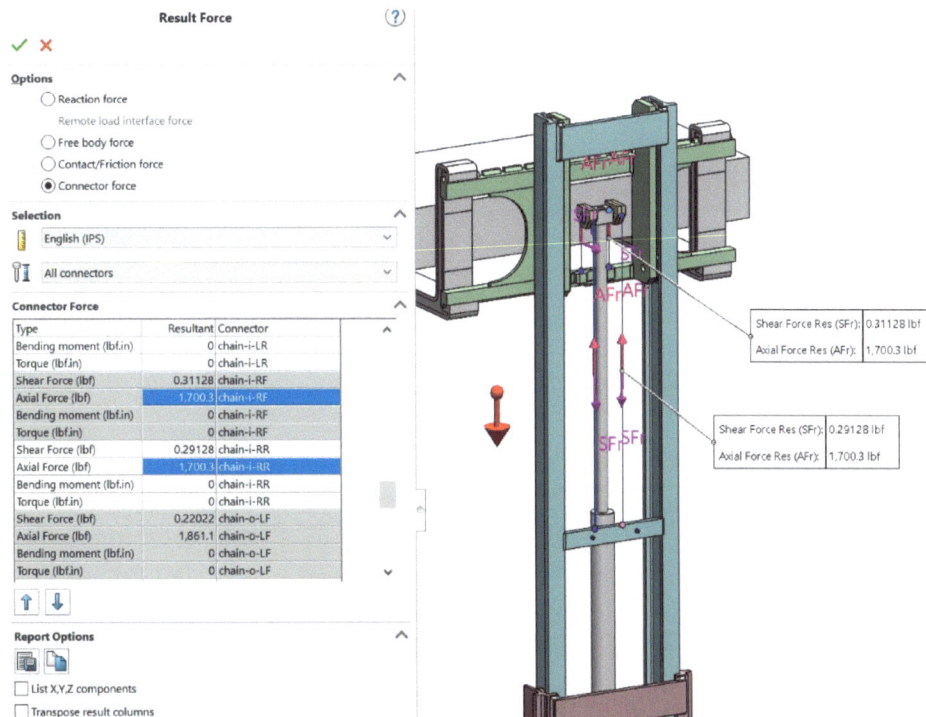

Result forces in selected chains

The "List Result Force" command is used to bring up the panel shown. Several forces in all types of connectors can be probed here, picked from the list and displayed in the graphics window. Here is where it pays off to name all the connectors.

We can be pretty sure that all the chains stay in tension in the vertical and forward lean studies. Forces there should be balanced left-to-right and the primary bending moment always works with gravity to keep the chains loaded. When the load and gravity are off-center, and the mast is tipped rearward, forces will be uneven left-to-right, and it may be that the chains on one side take all the load. From the result force screen a tabulation of all the forces can be output. With a little manipulation in a spreadsheet a clean table can be made to compare the results.

Connector	Type	1	2	3
chain-i-LF	Axial Force (lbf)	1703	1700	1735
chain-i-LR	Axial Force (lbf)	1703	1700	1735
chain-i-RF	Axial Force (lbf)	1700	1694	1647
chain-i-RR	Axial Force (lbf)	1700	1694	1647
chain-o-LF	Axial Force (lbf)	1861	1855	1288
chain-o-LR	Axial Force (lbf)	1861	1855	1288
chain-o-RF	Axial Force (lbf)	1860	1857	2410
chain-o-RR	Axial Force (lbf)	1860	1857	2410
Counterbore Screw-1	Axial Force (lbf)	1551	1143	1238
Counterbore Screw-1	Shear Force (lbf)	498	541	361
Counterbore Screw-2	Axial Force (lbf)	1565	1144	1062
Counterbore Screw-2	Shear Force (lbf)	499	542	953
Counterbore Screw-3	Axial Force (lbf)	2173	2018	2096
Counterbore Screw-3	Shear Force (lbf)	35	69	280
Counterbore Screw-4	Axial Force (lbf)	2266	2021	2088
Counterbore Screw-4	Shear Force (lbf)	35	69	279
Counterbore Screw-5	Axial Force (lbf)	2155	2014	2014
Counterbore Screw-5	Shear Force (lbf)	30	64	3
Counterbore Screw-6	Axial Force (lbf)	2293	2026	2049
Counterbore Screw-6	Shear Force (lbf)	29	63	2
cylinder-main-L	Axial Force (lbf)	-3887	-3877	-3863
cylinder-main-R	Axial Force (lbf)	-3887	-3877	-3863
cylinder-tilt-L	Axial Force (lbf)	2331	3354	783
cylinder-tilt-R	Axial Force (lbf)	2331	3354	783
sheave-i-L Joint 1	Shear Force (lbf)	3407	3402	3471
sheave-i-L Joint 2	Shear Force (lbf)	3407	3402	3471
sheave-i-R Joint 1	Shear Force (lbf)	3402	3390	3296
sheave-i-R Joint 2	Shear Force (lbf)	3402	3390	3296
sheave-o-L Joint 1	Shear Force (lbf)	3724	3711	2578
sheave-o-L Joint 2	Shear Force (lbf)	3724	3711	2578
sheave-o-R Joint 1	Shear Force (lbf)	3721	3715	4821
sheave-o-R Joint 2	Shear Force (lbf)	3721	3715	4821

Tabulation of result forces

There are eight segments of chain. The front and rear halves of each chain should be equal. In the first two studies we expect the left and right pairs of each connector type to be equal. This holds true to three digits consistently. The chains never go into compression, so we pass that check. In the third study more load goes to the right chain instead of the left (even though the load was biased to the left – this is why full assembly analysis is necessary for an assembly like a lift mast; counterintuitive quirks like this are missed in hand calculations).

The lift cylinders remain in compression, as they should, with a slight decrease in force when the mast is away from vertical. Force in the tilt cylinders increases as the mast tips forward and drops as the mast tilts back; the overall bending moment changes as the load block and mast weldments move further away from and closer to the truck axle.

Spring Chains

There is a connector type in Solidworks Simulation which can give us tension-only behavior. The spring connector can be set up may different ways, including "extension-only".

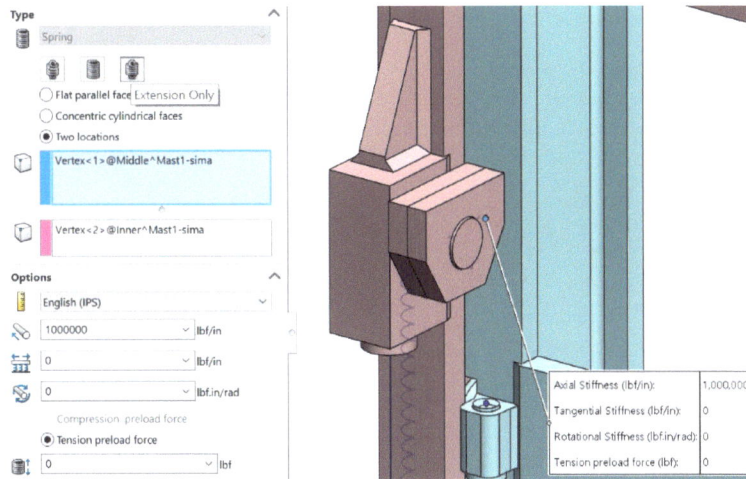

Extension-only spring definition

Here one of the rear chains is replaced with a spring connector. The extension-only type is picked, and the vertex-to-vertex option chosen. Now we can input a linear stiffness. Some leaf chain makers provide this data, per length of chain at each size. The stiffness input here will need to be calculated for each length of chain. It will be a smaller number for longer lengths of chain.

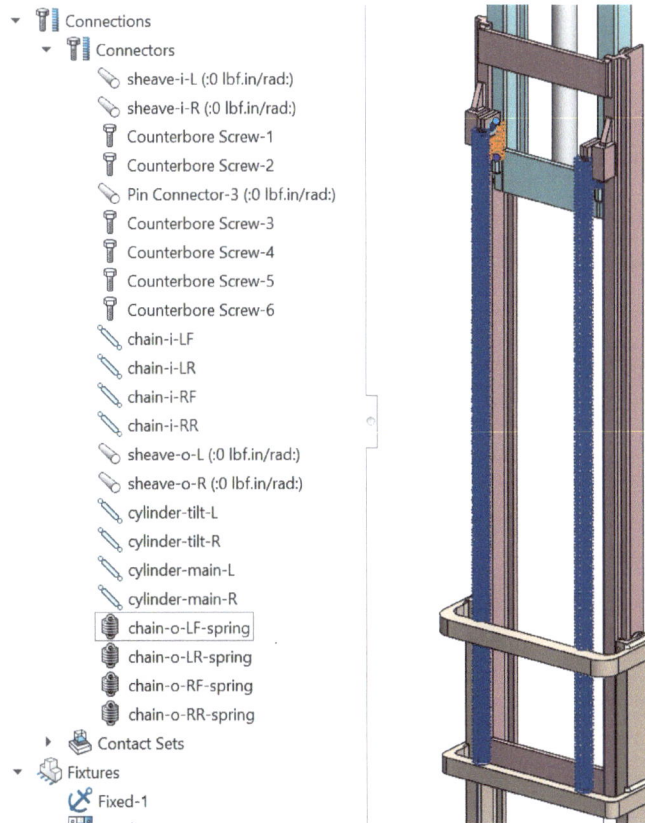

Connector set with spring main lift chains

Pictured is the new mast setup with all spring main lift chains. We have found that lift masts take much longer to run (and sometimes have trouble solving at all) with the extension-only spring chains. The solver must consider all possible outcomes of combinations of chains going slack. It may only be in unusual situations that a chain goes slack, but the simulation tool is there and experimentation with any new type of assembly is recommended.

Study 9: Casting Die Set

Overview

An aluminum housing is to be produced in a high pressure die cast process. The part inspired by a rear drive transmission bell housing but it much simpler.

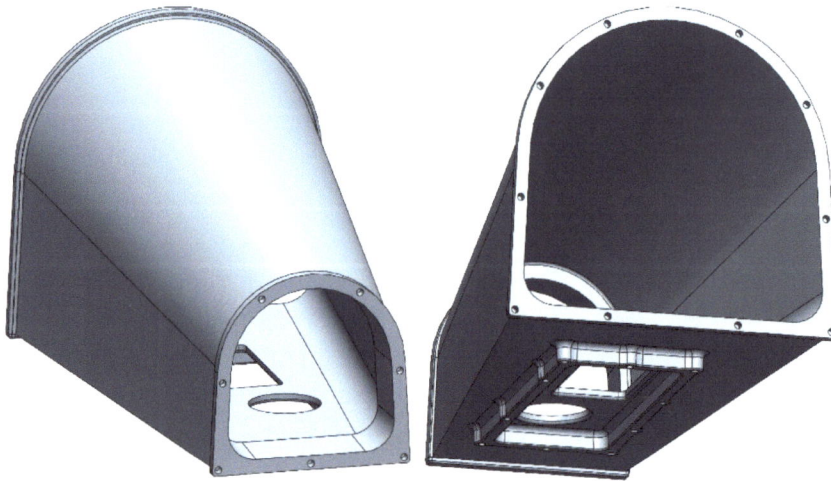

Model of part to be cast

The housing is 550 mm long, from the smaller flange to the larger end. The internal shape is made by two moving die pieces, or slides. External shapes in the middle are drafted so as to be producible directly in the main die inserts.

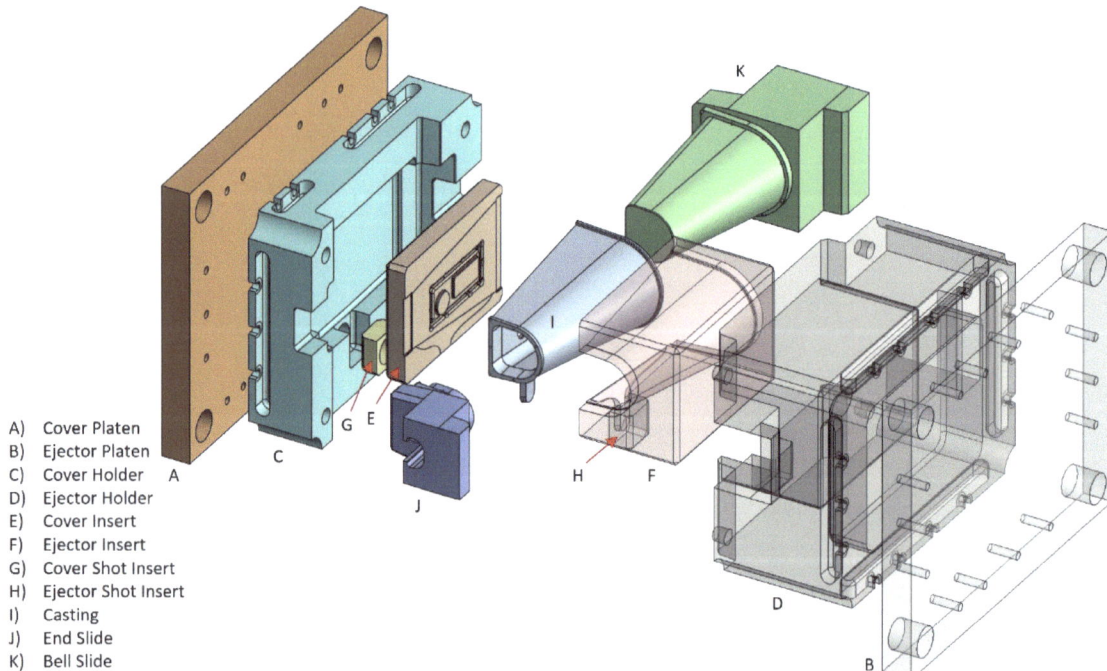

A) Cover Platen
B) Ejector Platen
C) Cover Holder
D) Ejector Holder
E) Cover Insert
F) Ejector Insert
G) Cover Shot Insert
H) Ejector Shot Insert
I) Casting
J) End Slide
K) Bell Slide

Exploded view of die set

The planned die set is shown in an exploded view. These parts assemble into a diecasting machine which carries the two platens on four tie bars (not shown). Liquid aluminum is pushed into the cavity from the "cover" side. The "ejector" side parts (transparent) move together on the tie bars. The die cast machine can press the platens together with 2000 short tons (4 million pounds) of force, while the cast aluminum is pushed into the cavity at 12,000 psi. The die pieces all together come to nearly 12,000 pounds mass of high alloy steel.

Objective

We want to know if the machine size is adequate, and if simple retention of the slides will keep them in position under pressure. Stresses in the main parts are of interest so that large cooling lines are not cut through high stress areas. We also want to check stresses in the case of an unplanned interference, like from a chip of material or build up of flash being trapped between parts.

Challenges

The key to this job will be very intensive contact analysis. Also some of the main parts are bolted together and this may need to be included. Very long run times are expected.

The Die Model

At this tooling design stage the die model consists of just ten tight-fit blocks. A section view through the shot (metal inlet) shows how the inserts fit tightly in the holder blocks. A single cross hole at the top is where one large cooling channel is planned.

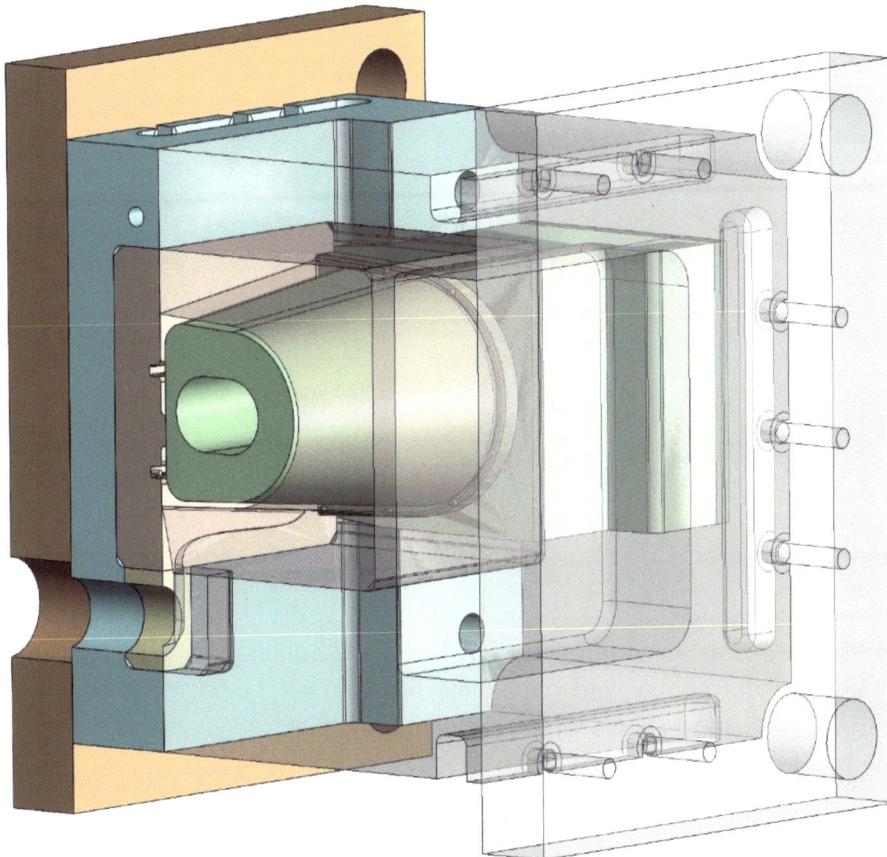

Section view through die set, vertical

Another section through the slides shows how they are meant to fit.

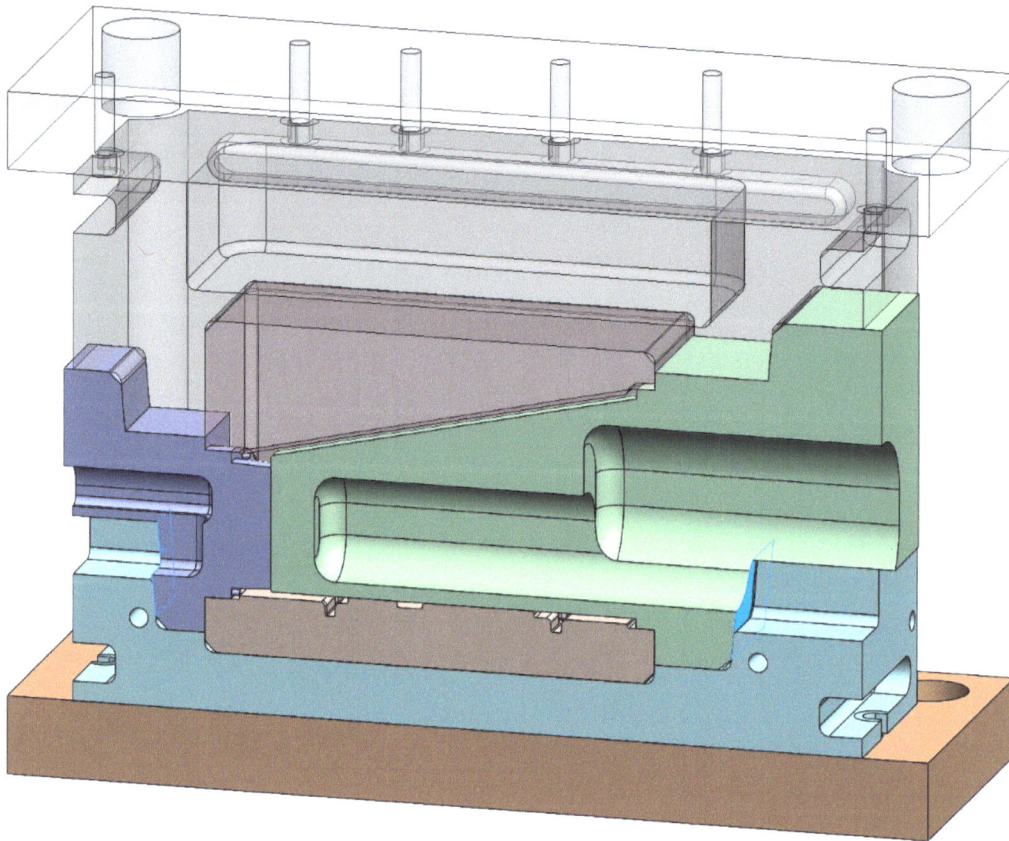

Section view through die set, horizontal

The slides move with the ejector side. They are pulled out before the cast part is extracted, then cleaned and slid back into position before the die closes again. In reality a finished die set will have many more mating parts, as shimmable pads are installed which are adjusted to the final fit of all the contacting parts. But the intended end result is a set of moving parts which come together in a near perfect close fit.

A sloped surface on each slide (highlighted) contacts the side walls of the cover holder and is the primary lock of the slide against injection pressure. The cover holder is necessarily cut away some in these areas, to make room for cooling and mechanical hardware in each slide, so stress in a cover holder is sometimes an issue. Here another two large cooling lines can be seen running between the lock faces and the mounting bolt locations.

First Run, Cover Holder Only

We see already that the cover holder is a part of key interest. Previous experience, including some real broken dies, backs this up. So we begin with just that part.

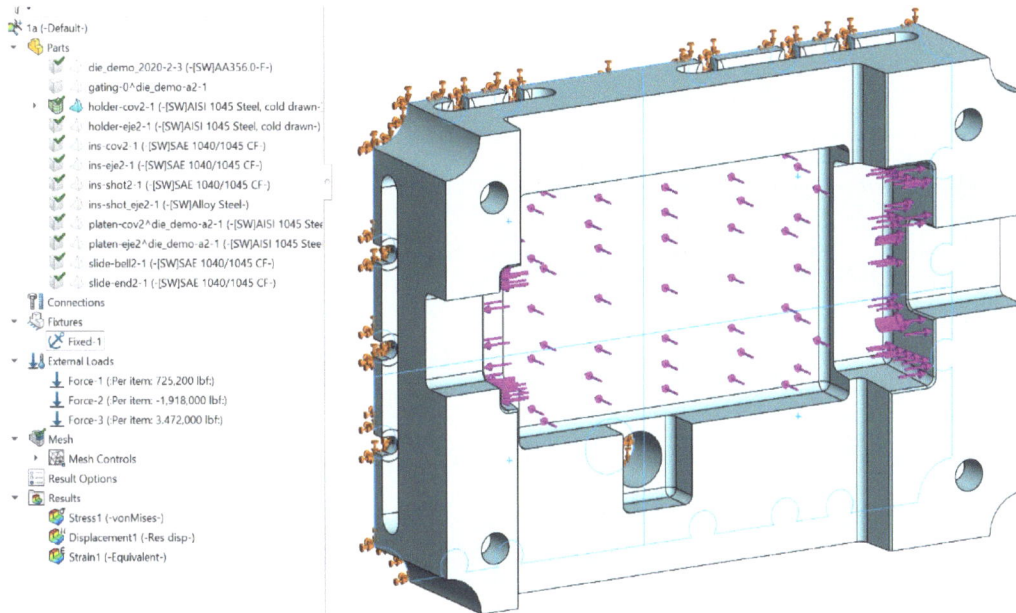

Initial setup, cover holder with fixed face

Resultant forces are estimated from the projected area of the cavity in each direction and the planned fluid pressure. These forces are applied to the respective faces of the cover holder. Each slide lock face is assumed to take all the push from its respective slide. The entire back face of the holder is fixed.

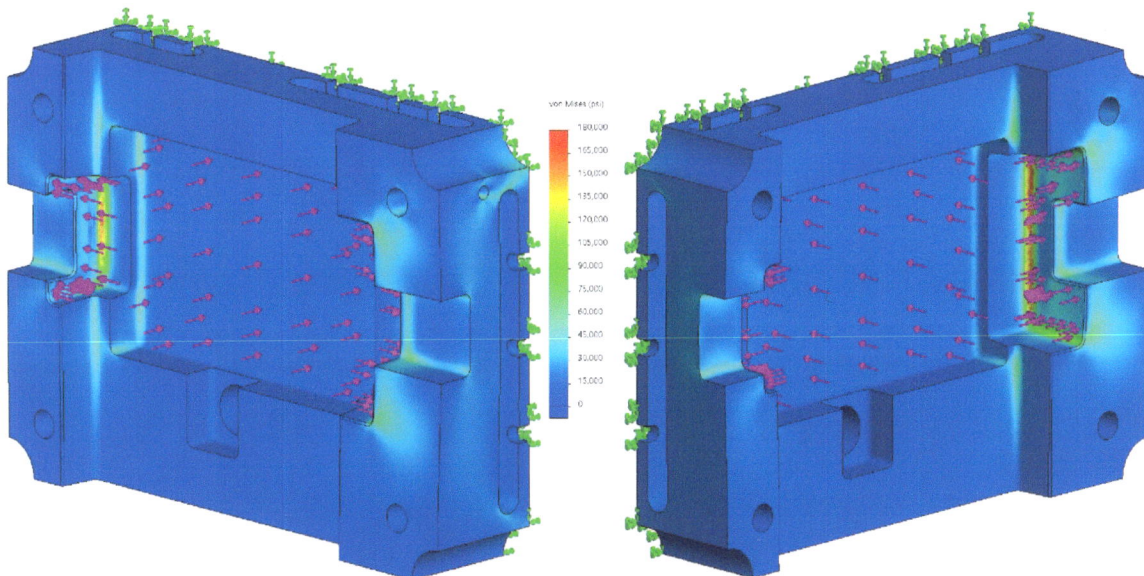

Stress on cover holder

The result shows some stress concentrations at the inside radius bordering each lock face. The mesh isn't shown; it's pretty coarse but the curvature-based mesher gave us good refinement on all those inside radii.

Cover Holder, Bolted

The result is "reasonable" but we know the large surface fixture is not right. A quick improvement can be had using foundation bolts.

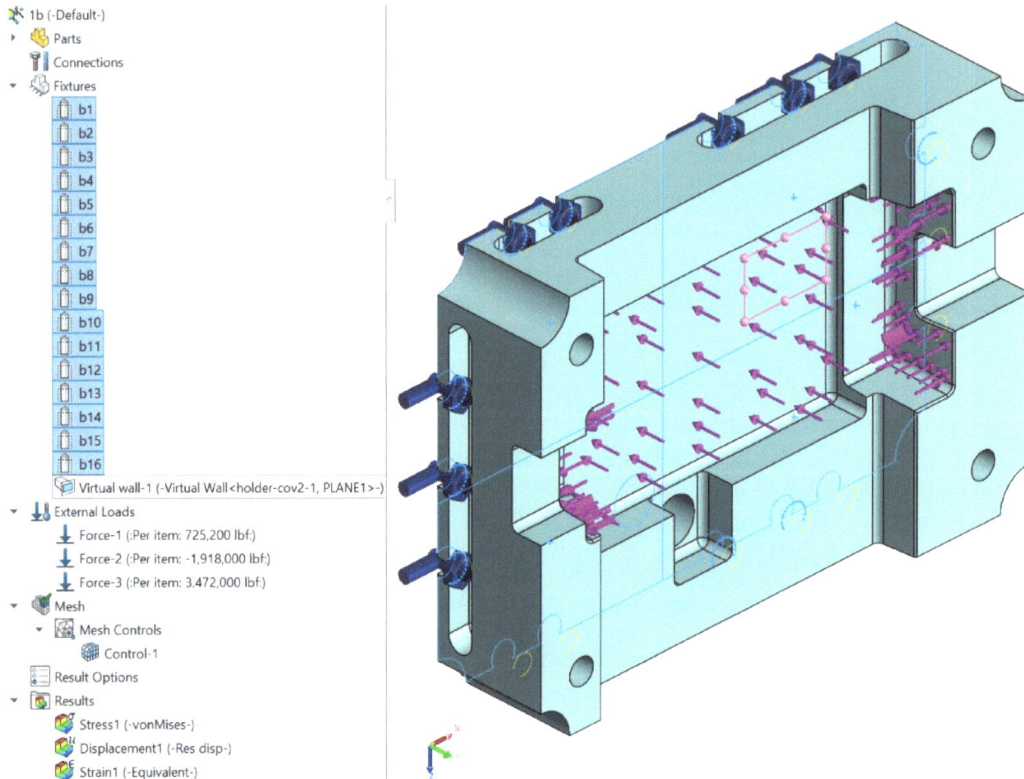

Setup with foundation bolts

Running with the same forces the surface stress result is just a little hotter. Section views illustrate how the bolt fixtures interact with the applied forces.

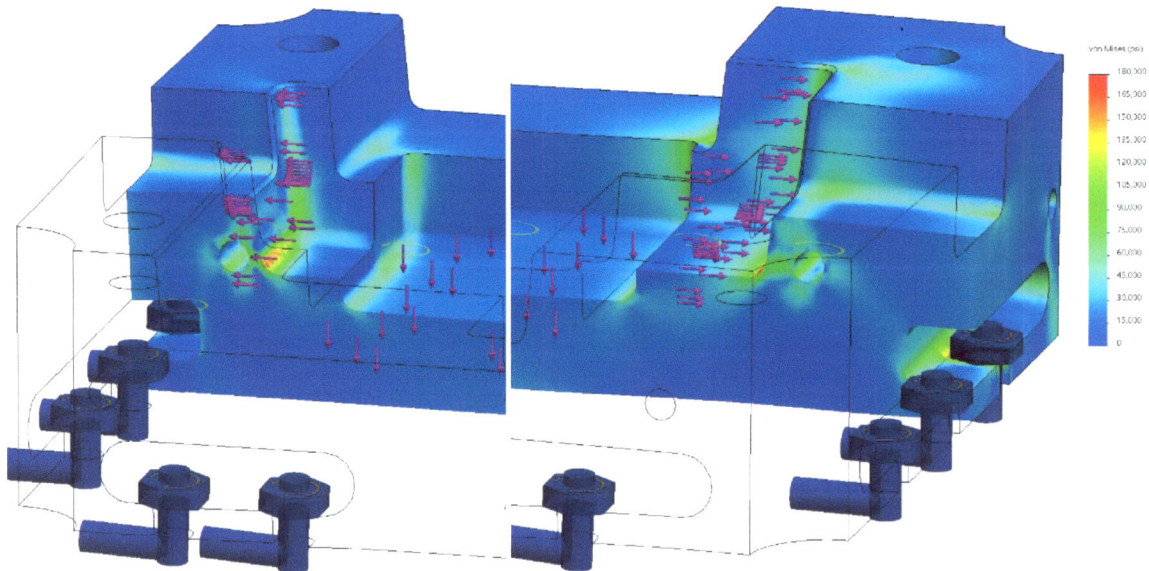

Section through stress result (15x displacement)

Still we know that applying forces directly to faces of an object is not realistic. There will be solid bodies pushing against this holder block, and they will concentrate force differently than a uniform force application.

Force Through Pusher Parts

We bring some of the other die parts back into the model. The same forces are now applied to the slides and cover insert, which are put into contact with the cover holder.

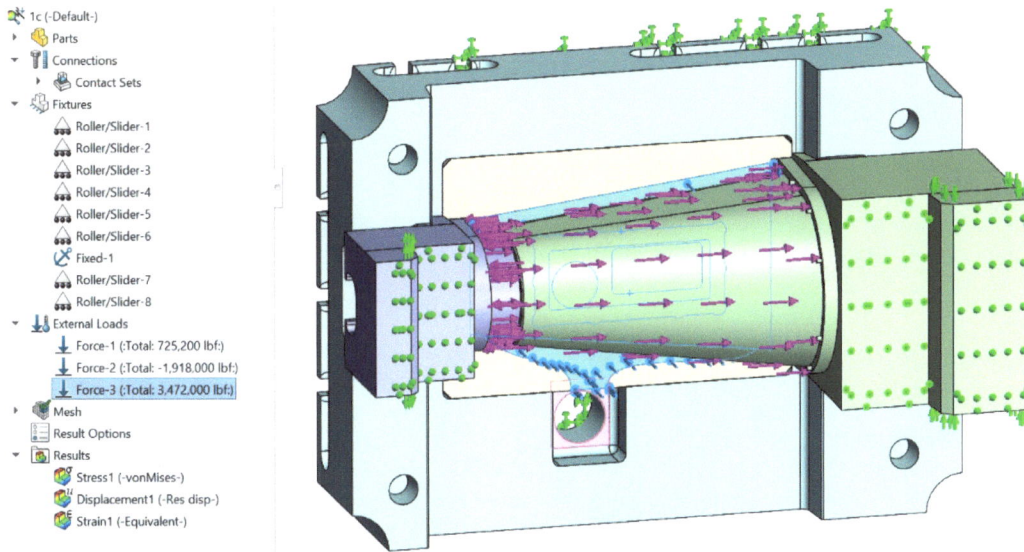

Setup with pusher blocks

Planar roller/slider fixtures lock the slides into only left-right motion. For speed we've gone back to a fixed plane behind the cover holder. In these stress plots we hide the new blocks and some of the simulation symbols.

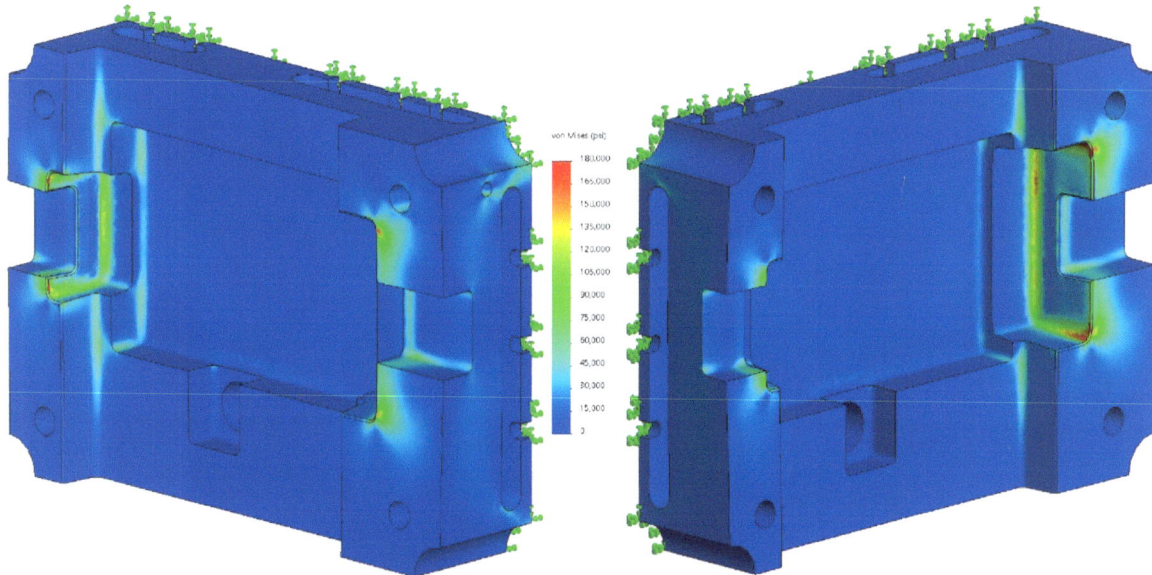

Stress result from load application through pusher parts

The stress result differs from the first run. Peaks in the deeper radii are lower, but there are contact edges that show up clearly at the near surfaces.

Pusher Parts and Bolted Holder

To be complete we put it all together. The holder block is fixed with foundation bolts against a virtual plane.

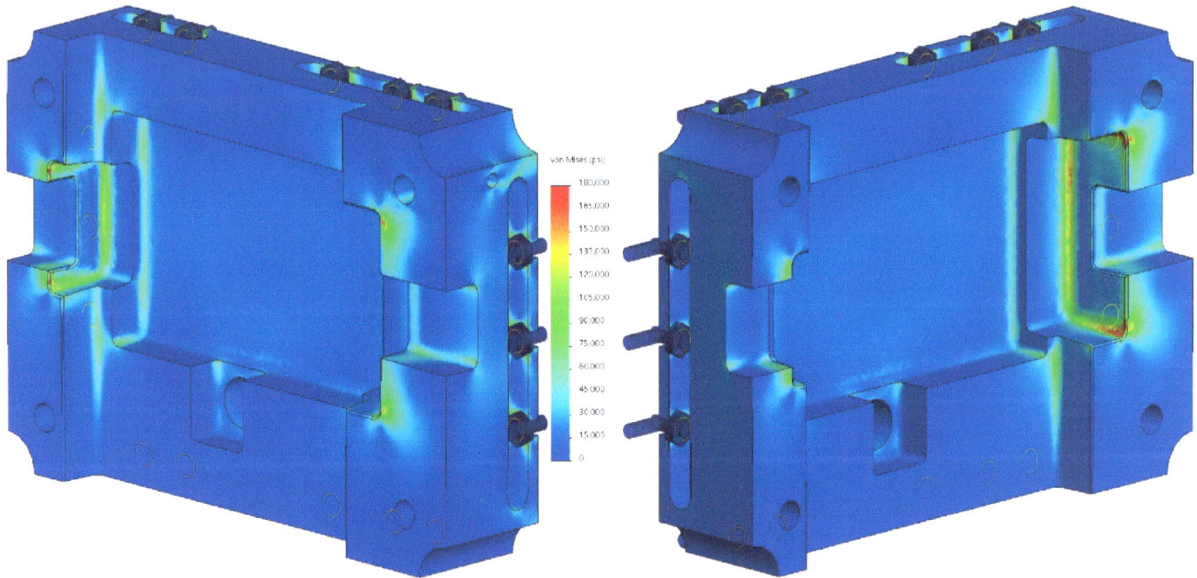

Stress result with pusher parts and bolted base

The surface result doesn't change much. A section view, with all parts shown, reveals more of the interactions. Already we can see the slides wanting to buck upward in reaction to the sloped lock face forces. Some interesting stress fields are seen around the big cooling holes.

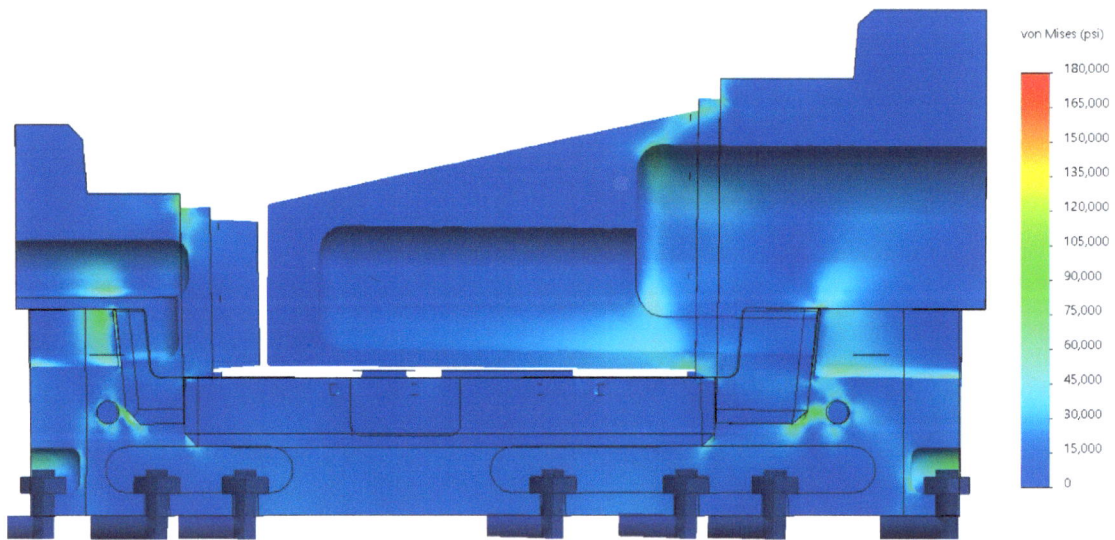

Section view through complete stress result (5X displacement)

We are getting more interesting, and we expect more usable, results, but at what cost? The model size did not grow much with the added pusher parts, but the contact and bolts do cost something as we've seen before. Run time on the fourth study is almost 38 times longer than the first single-part study. But that first study only took 74 seconds.

Running Both Sides

The cover holder does not work alone of course. Clamping force from the ejector holder may affect it materially. So we add the other holder back in to the model.

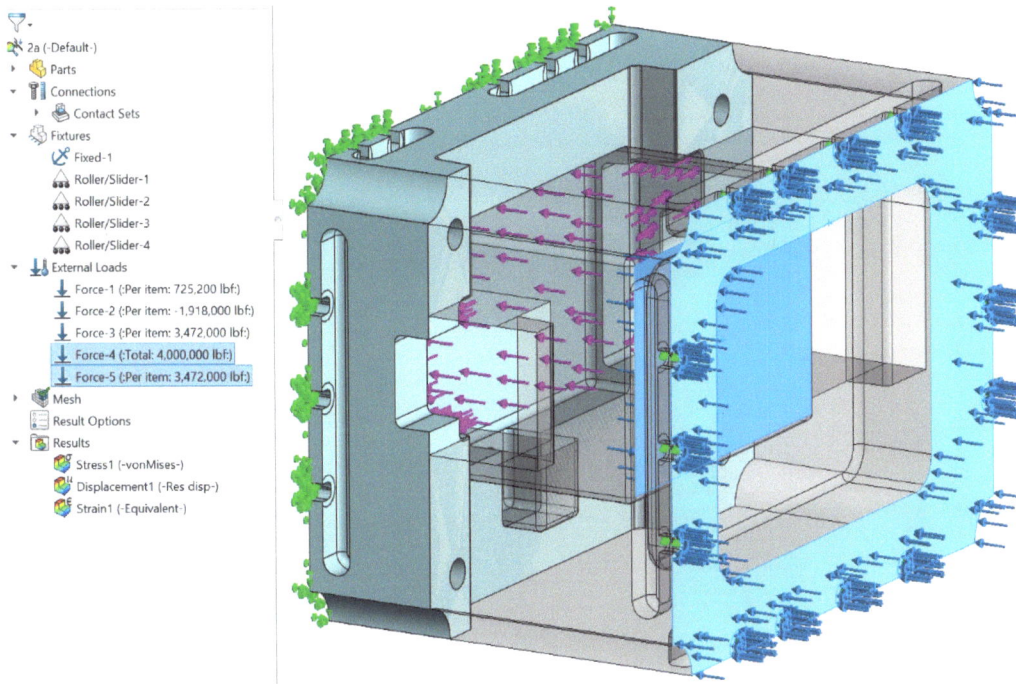

Setup with ejector holder added

The cover holder fixture is back to a simple fixed plane. Expected fluid pressure forces are applied to the cover holder as before. But now the ejector holder is pressing against it with the planned holding force of 4 million pounds, minus the same fluid pressure force as the cover side cavity sees in the opening direction. For stability roller/slider fixtures are put on small faces inside each ejector holder bolt slot.

Fixtures in bolt slots

Contact was manually defined between all intimate surface pairs of the holders, 9 sets in total. The initial result is dramatic.

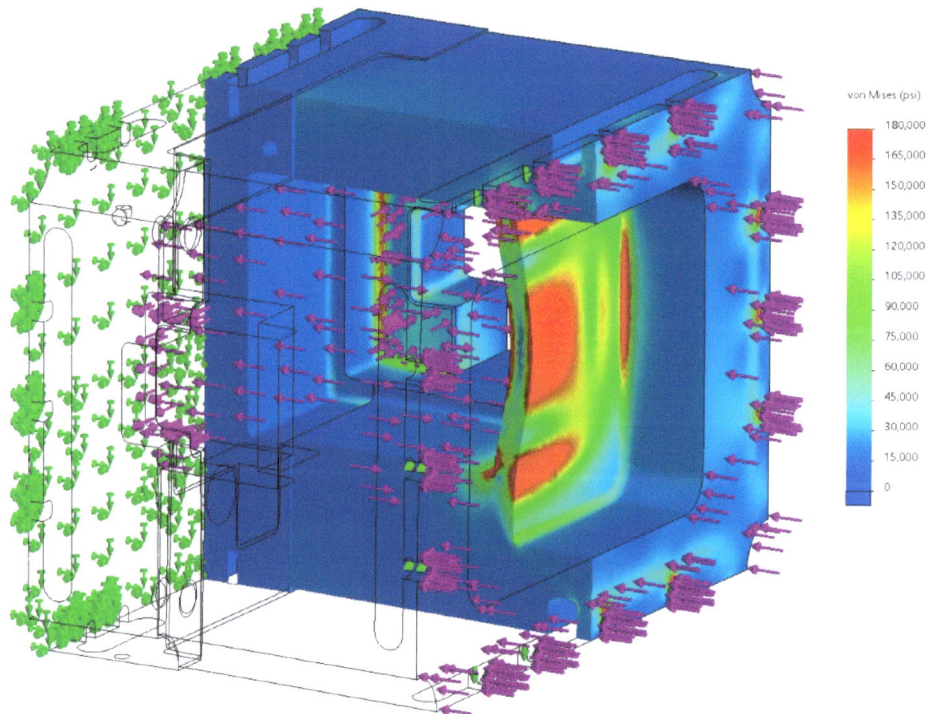

Section through stress result (5X displacement)

The ejector side holder has a large cavity in the middle. In this area ejection hardware is typically installed in addition to fittings for cooling lines and other hardware. With a 3.5 million pound load applied to the cavity side the remaining relatively thin (38 mm) wall is pushed outward and highly stressed.

The result on the ejector holder is not changed by converting the fixture again to foundation bolts.

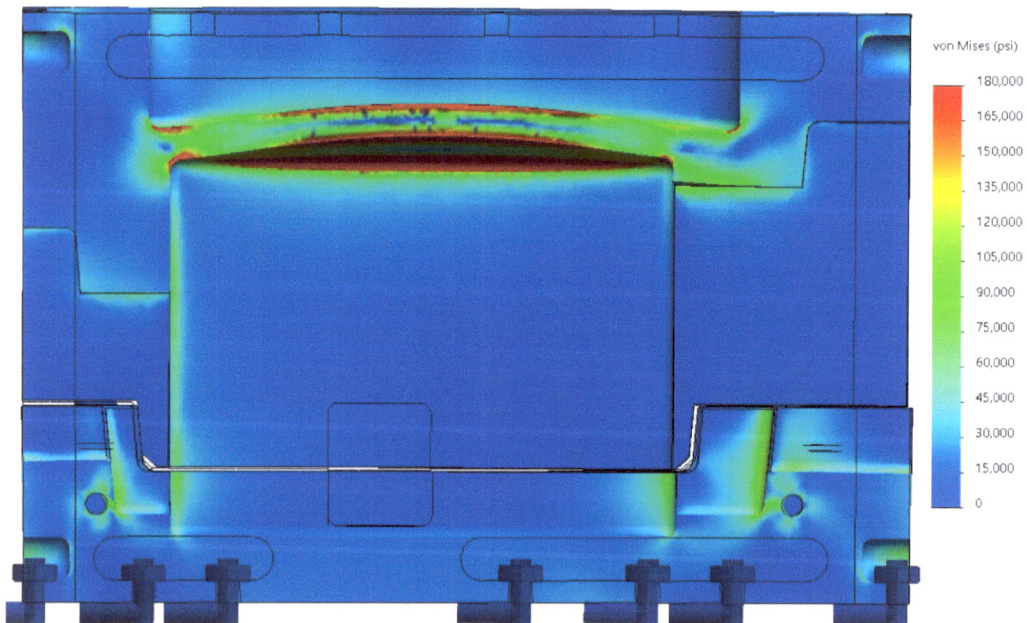

Section through stress results (5X displacement)

The result in the cover side may show subtle differences to the first runs but mostly we see the ejector side trying to lift off, unevenly, in addition to its central distortion.

Including All Die Parts

The ultimate goal here was to run the complete die set, in realistic contact. All the die parts are added back in to the model.

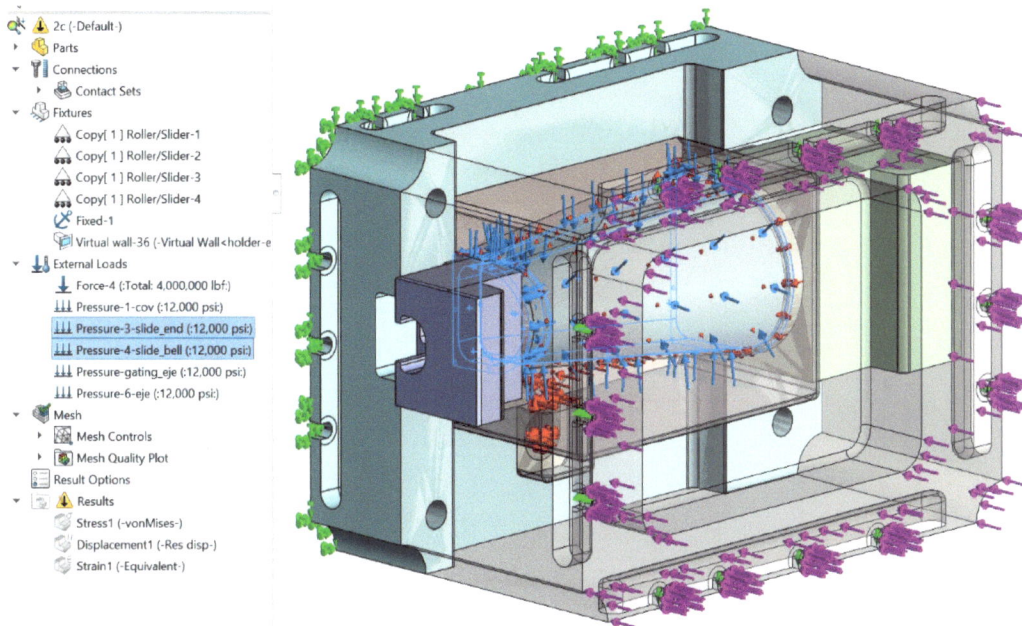

Setup with all components included

The manually applied forces are removed. Now that the real part geometry (in negative space) is in the model, we can apply fluid pressure to those surfaces. Total force vectors will be computed by the computer instead of the user. The setup is easier to visualize (and define) on an exploded view.

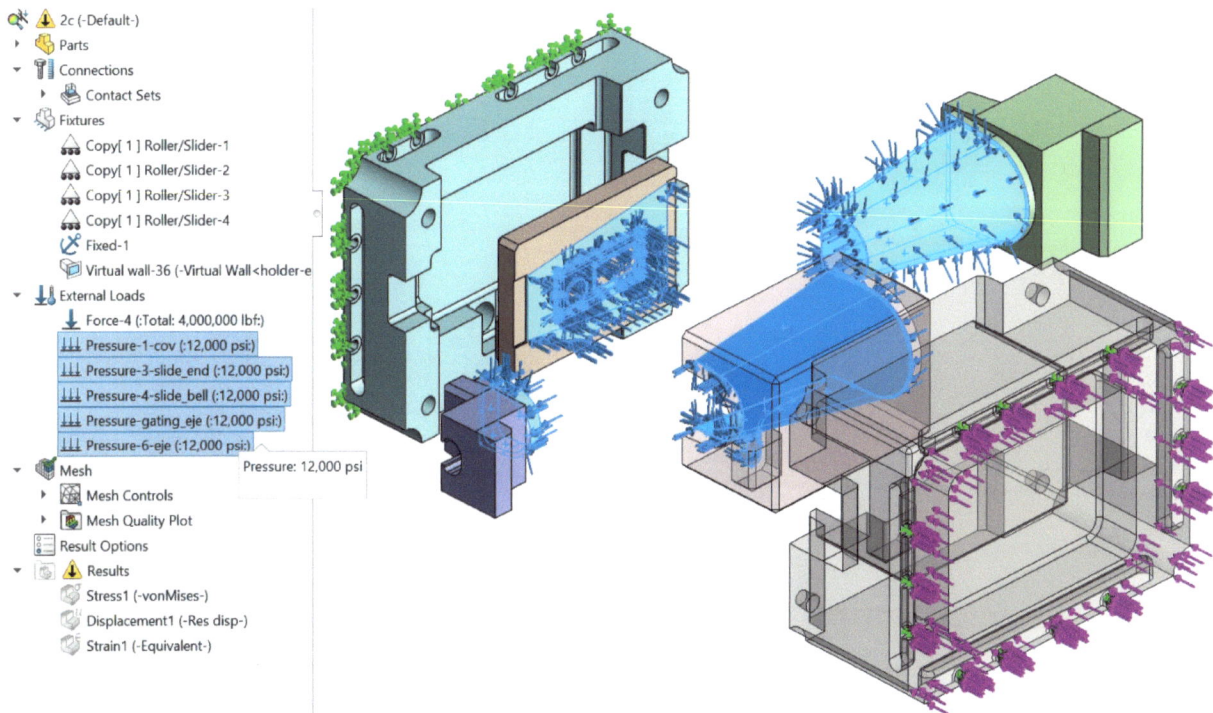

Setup in exploded view, pressure conditions highlighted

The result is radically different from when forces were applied directly to the holders.

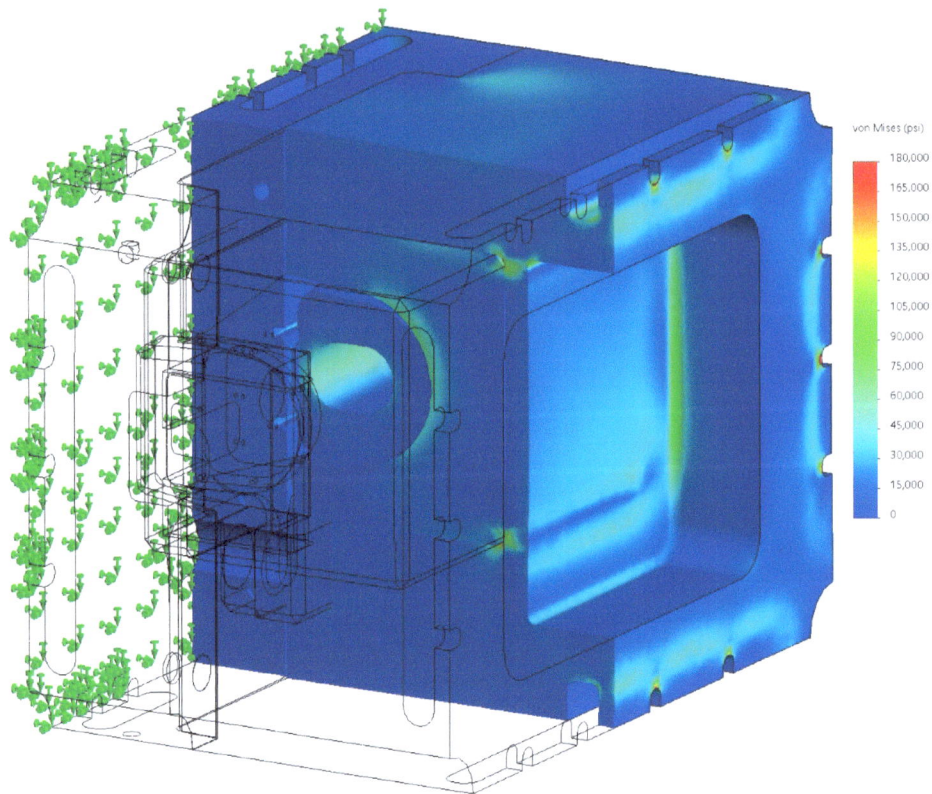

Section through result with all components (5X displacement)

The center of the thin part of the ejector holder sees much less of the load with the main insert pushing into it. The same result is seen on the familiar exploded view.

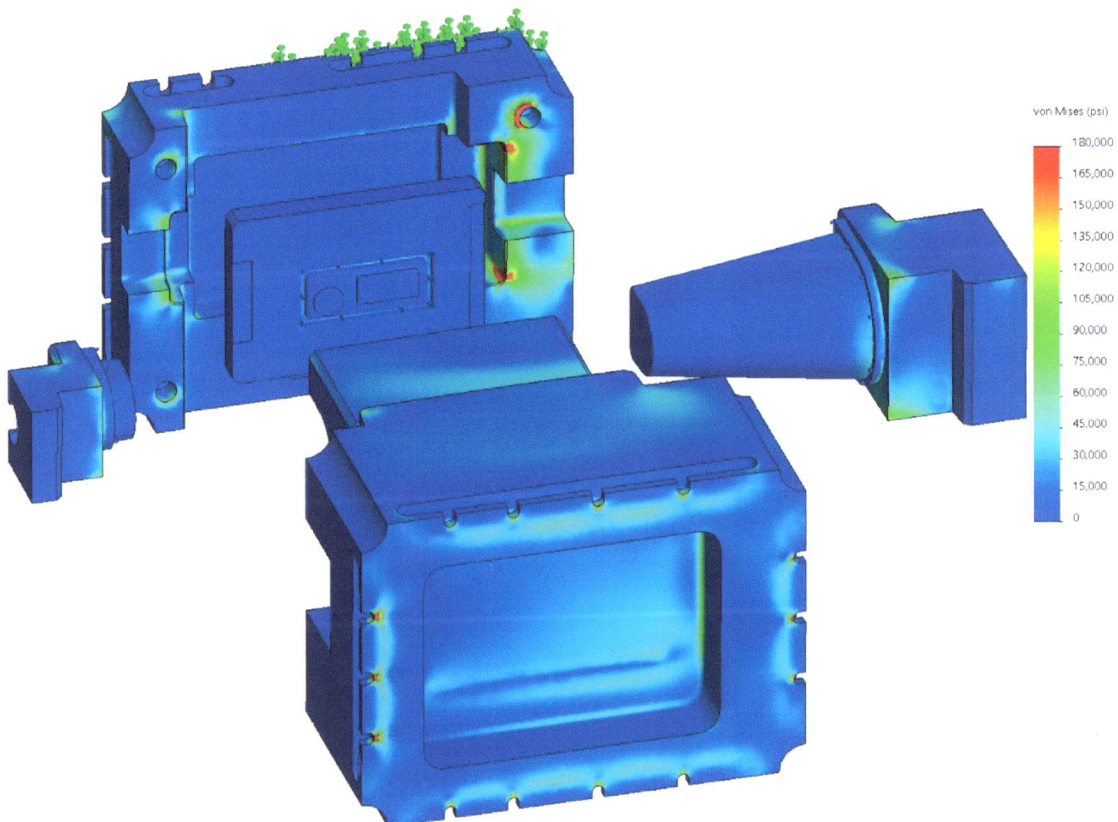

Result on exploded view

The Real Deal

Before we look at the result in more detail, we note that we still are fixing the parts that matter to us, the die holders. There is one more step to get to a truly "live" machine which moves like the real thing and doesn't artificially constrain the parts of interest. The die cast machine platens are added to the model, joined by large pin connectors.

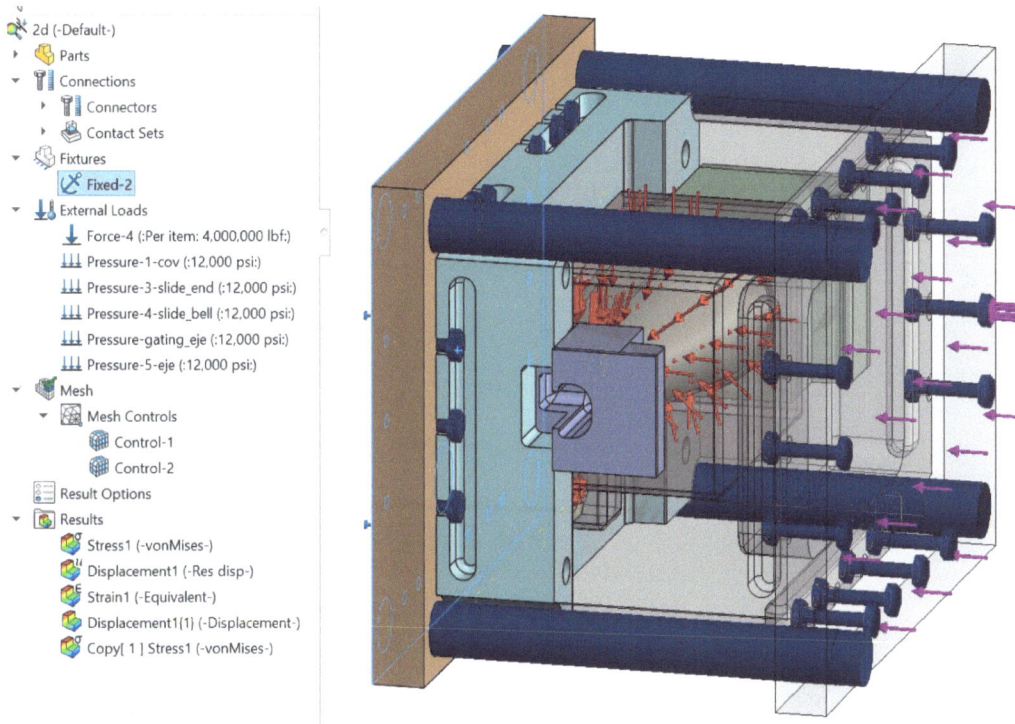

Setup including machine platens

The back side of the cover platen is fixed. Each die holder is bolted to its platen. The ejector side platen is free to slide on the four pins. The 4 million pound holding force is applied to the ejector side platen. Fluid pressure is the only other input force.

In this setup we have every bit of realistic flexibility that can be put into this model. It should behave just like the real deal down to a close level of detail. With all these parts in contact and dozens of bolt connectors it runs for a very long time (over 6 hours on our reference workstation) and an error pops up, "Equilibrium is not achieved"[2]. The error is dismissed by accepting the last iteration. Looking at the imperfect result is sometimes a pretty direct way to debug this.

---

[2] Inspection of the .OUT file will show that the number of contact force iterations reached 100

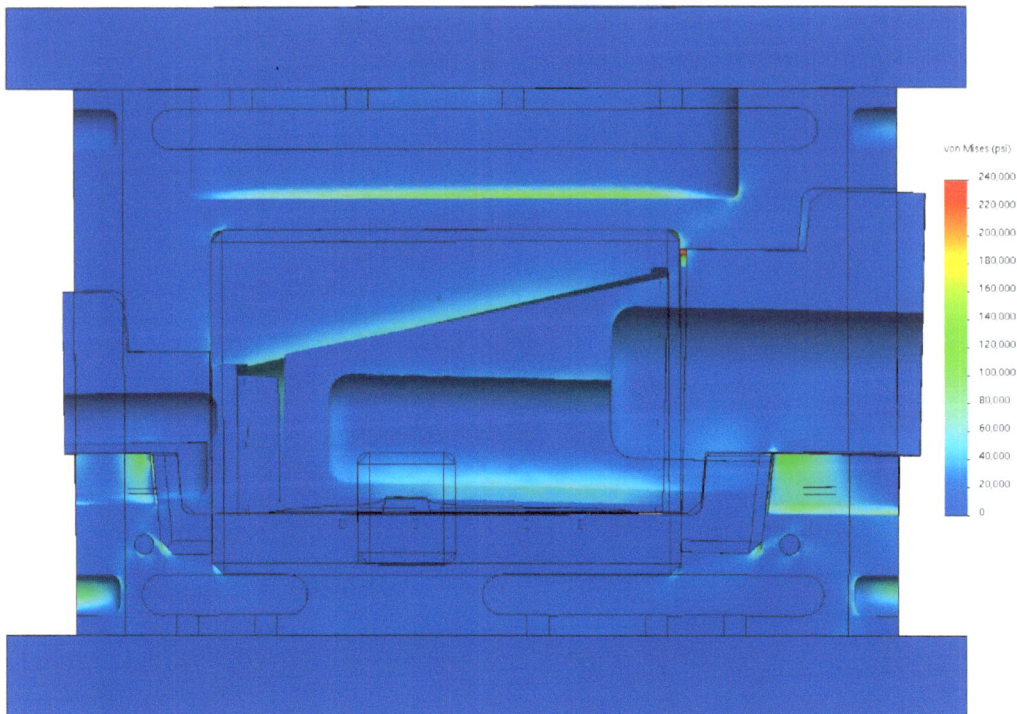

Section through result with platens (2X displacement)

With even small displacement magnification the severe kick of the slides can be seen. The larger slide will push into the ejector side die insert, leaving no thickness in the cast part. In addition there are large inter-holder forces and resultant stresses where alignment pins connect the holders.

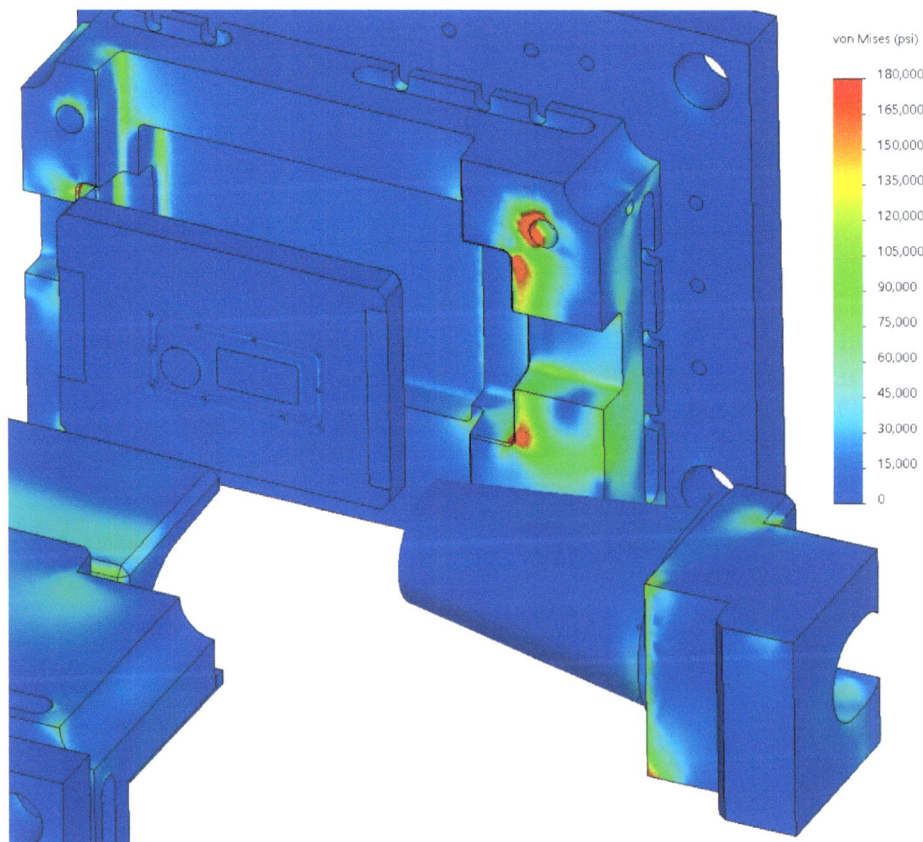

Result on exploded view

We could put more work into getting a fast and resolved run, but a simple design change is implemented which is expected to help both the mechanical and numerical stability of the assembly.

Added Rails
Rails and keyways are added between the slides and the ejector side holder block.

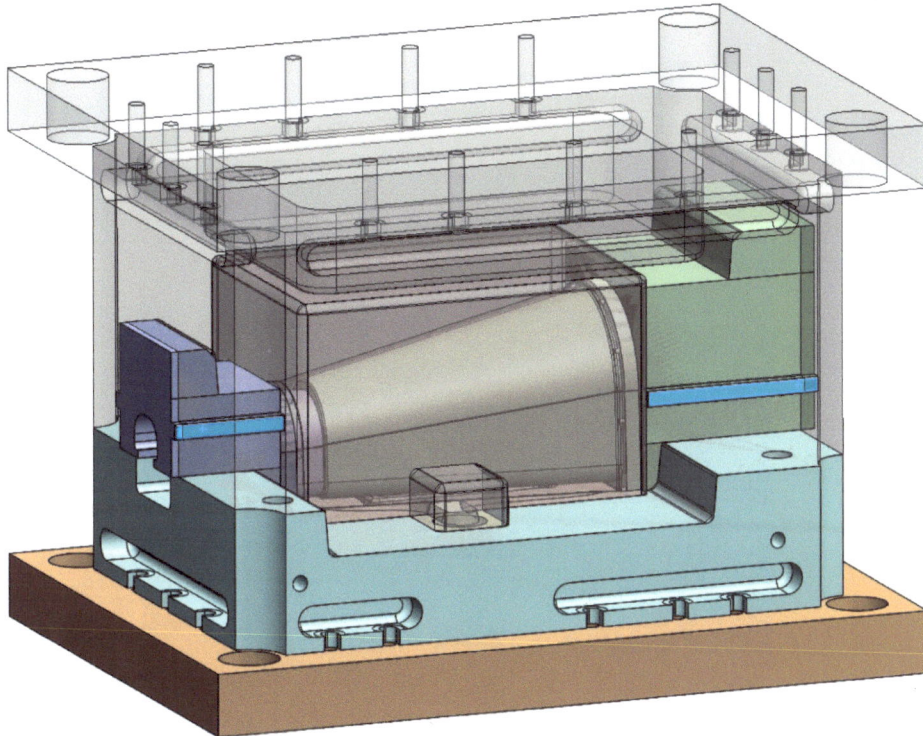

Model with rails on slides

The rails are expected to keep the slides from rotating out of position when loaded against the angled lock faces. Even though they added several more contact sets it is also expected that they will help the simulation resolve and complete faster.

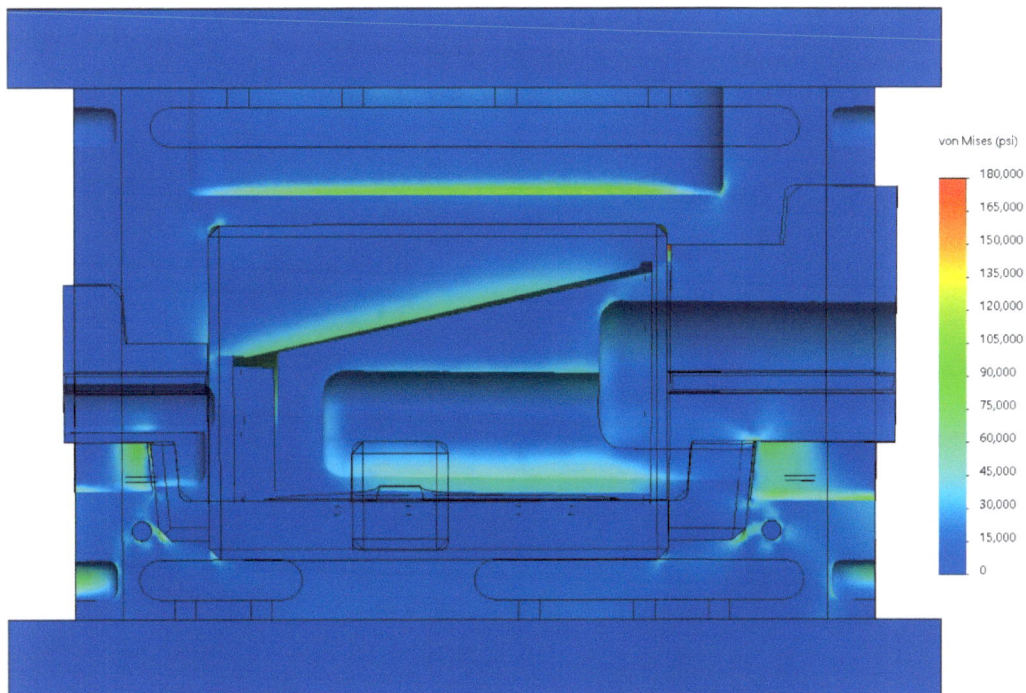

Section through result with added rails (2X displacement)

The run does finish faster and without warning messages. Mechanical stability is clearly improved, though the die designer still have work to do keeping the large slide in location [this is a known design challenge with large slides; bringing support higher up from the cover holder can stabilize the slide but it makes for tall unsupported towers in the holder]. We can pull a variety of stress plots and other data from this result.

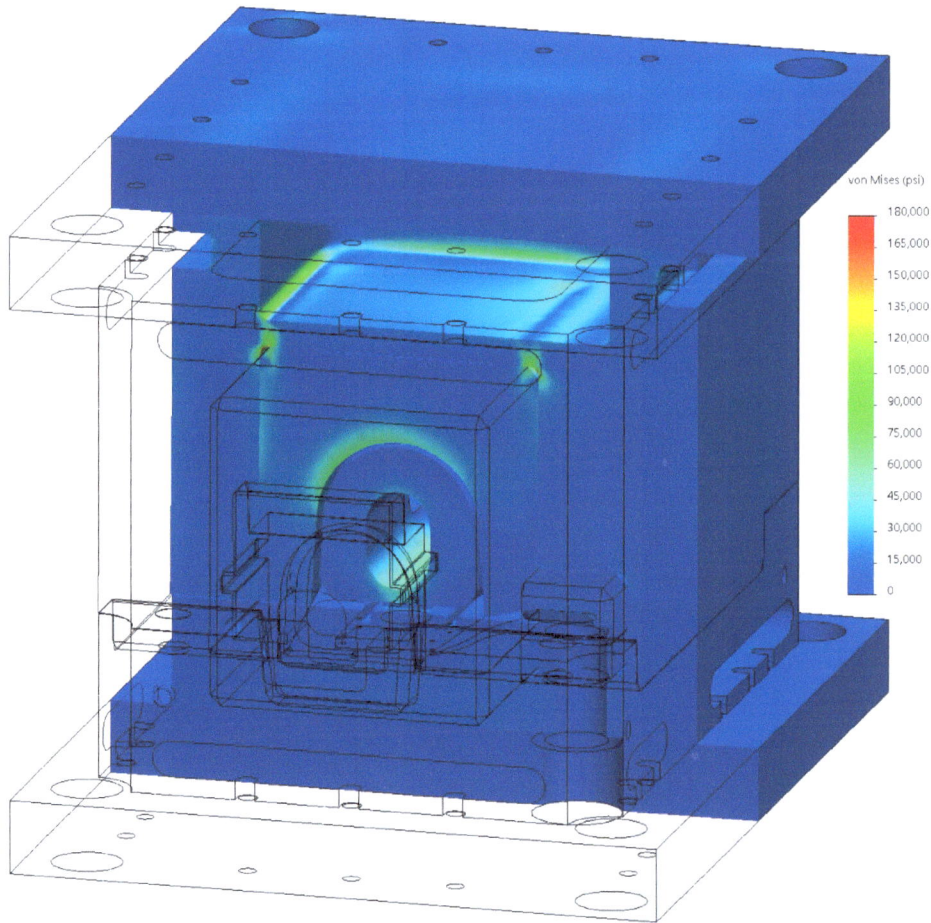

Section through result with added rails (2X displacement)

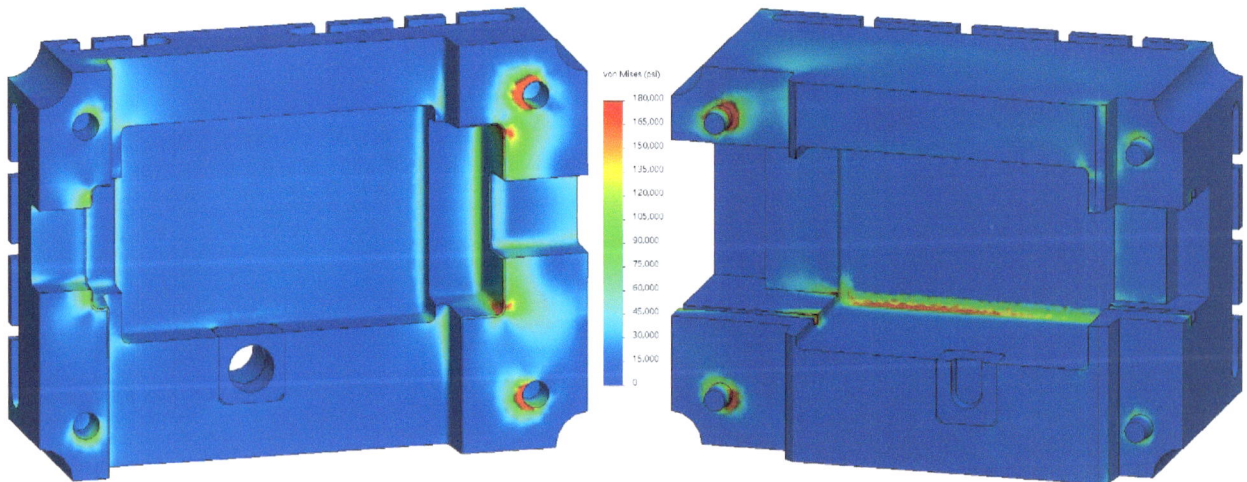

Surface stress on holder blocks

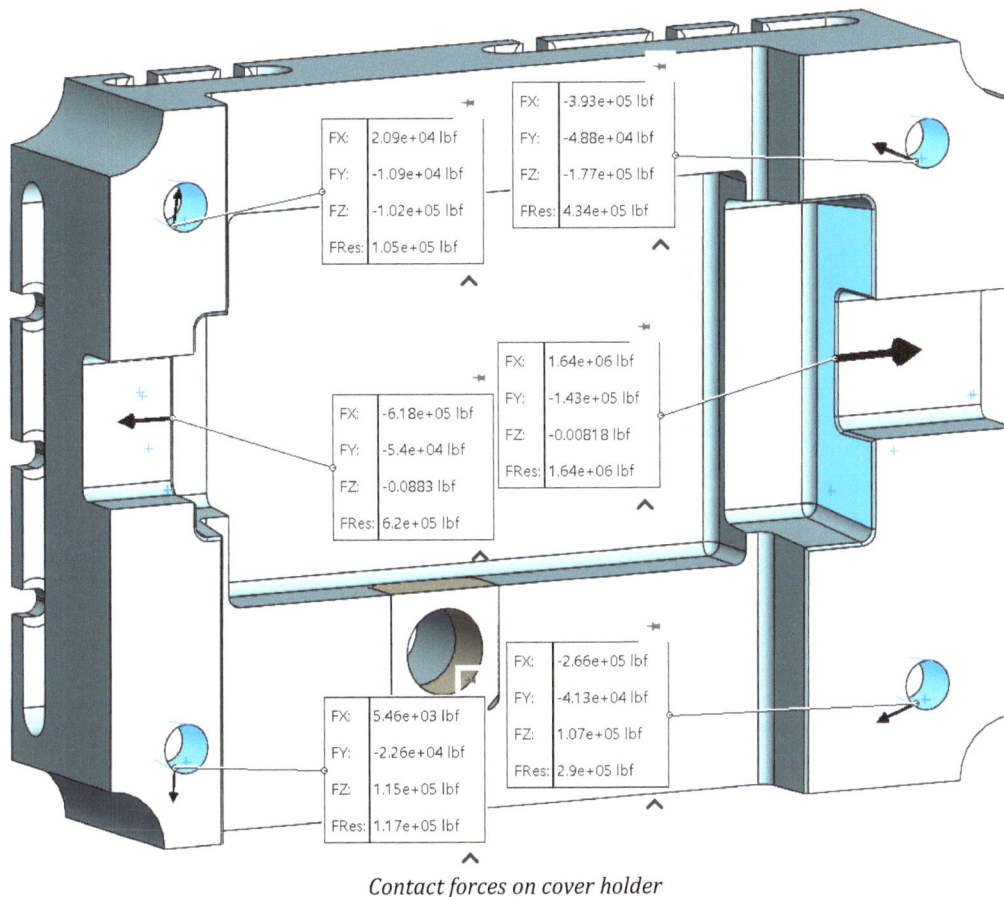

FX:	2.09e+04 lbf
FY:	-1.09e+04 lbf
FZ:	-1.02e+05 lbf
FRes:	1.05e+05 lbf

FX:	-3.93e+05 lbf
FY:	-4.88e+04 lbf
FZ:	-1.77e+05 lbf
FRes:	4.34e+05 lbf

FX:	1.64e+06 lbf
FY:	-1.43e+05 lbf
FZ:	-0.00818 lbf
FRes:	1.64e+06 lbf

FX:	-6.18e+05 lbf
FY:	-5.4e+04 lbf
FZ:	-0.0883 lbf
FRes:	6.2e+05 lbf

FX:	-2.66e+05 lbf
FY:	-4.13e+04 lbf
FZ:	1.07e+05 lbf
FRes:	2.9e+05 lbf

FX:	5.46e+03 lbf
FY:	-2.26e+04 lbf
FZ:	1.15e+05 lbf
FRes:	1.17e+05 lbf

Contact forces on cover holder

In general, we find that the die cast machine has enough clamping force (or none of the later runs would have completed – the ejector would have been pushed away). But the platens may be under sized in dimension. The die designer will struggle to find enough steel around the outside of the die inserts to rigidly back up the substantial fluid pressure in the cavity.

Unplanned Interference

We are not quite done with this die set. As modeled all the parts are perfectly matched, snug fitting and flat at every contact. But a real die has a lot to put up with, temperature extremes which degrade surfaces and debris which can get lodged into those tight clearances. Stone Lake has some experience with real broken dies and here we look at an example of hard chips being caught in the worst possible places.

Debris in a casting die can come in different forms. The most likely material is the alloy being molded, in this case a common aluminum. Flash is known to build up in layers sometimes before being dislodged and not always blown away. For our purposes we will make an interference in steel by adding to one of the parts. We are not looking for an exact solution near the interference, this would require an intensive nonlinear analysis. A thin interference in steel will be our surrogate for a thicker bit of rogue aluminum.

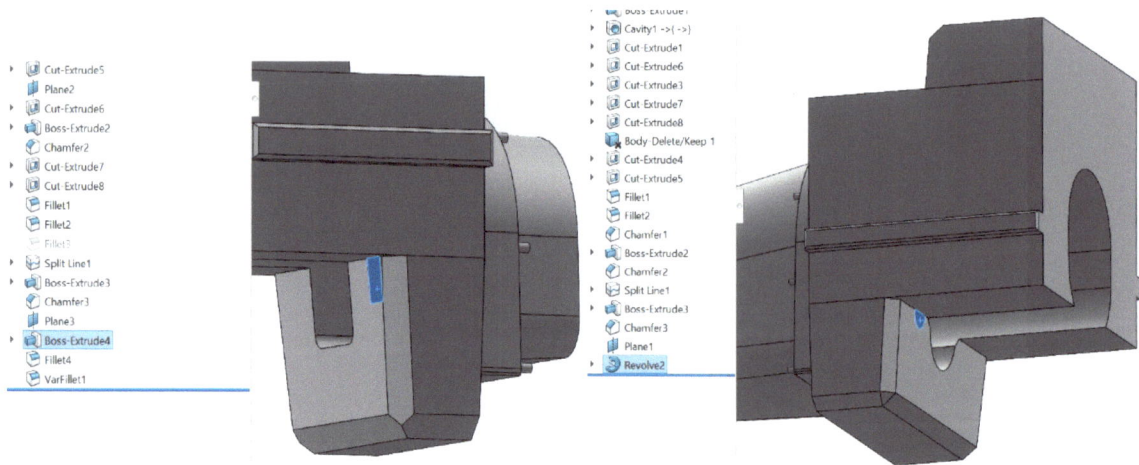

Modified slide models

The bell end slide is modified with a revolve operation that makes a thin blister on the lock face. The smaller slide is similarly modified but with an extrude operation to make a wedge-shaped problem. The exact size and shape is a matter for discussion with the process engineers. We want to look at the bulk effect on robustness of the die holders.

The new blisters are put in contact with split surfaces on the cover holder using the "Shrink Fit" contact set type. The mesh is refined liberally on both sides.

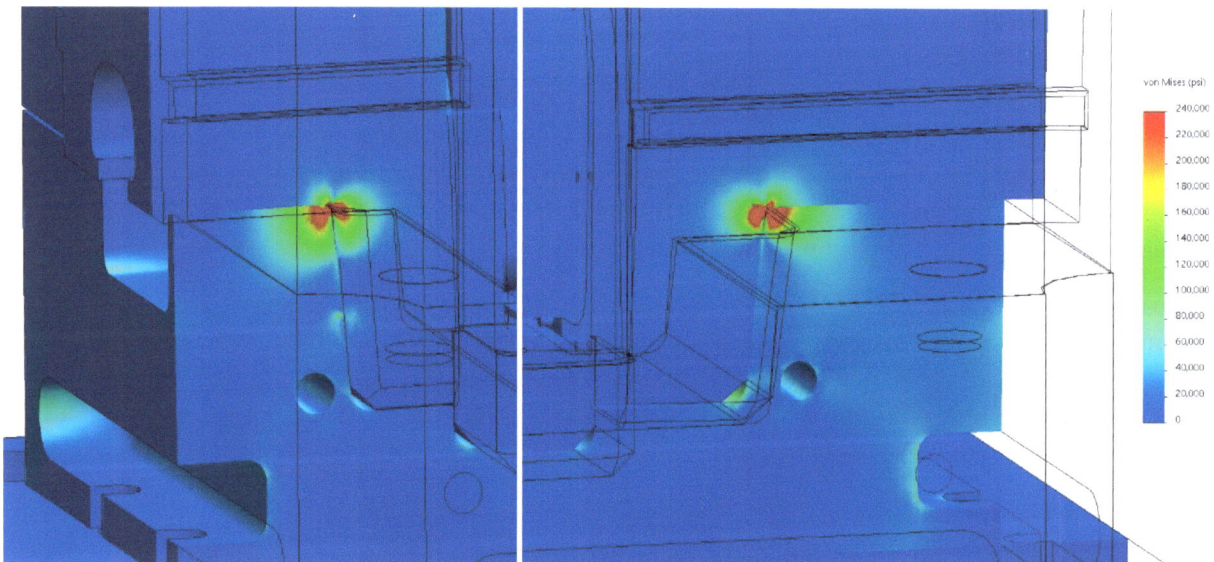

Section through results at new interferences

Extreme stresses are seen where expected at the interferences. Again, we will not attempt to compute true stresses at the contact patch. What we want to see here in particular is if the planned cooling passages cause any trouble. Elevated stress can be seen around both holes, but not to any troublesome level.

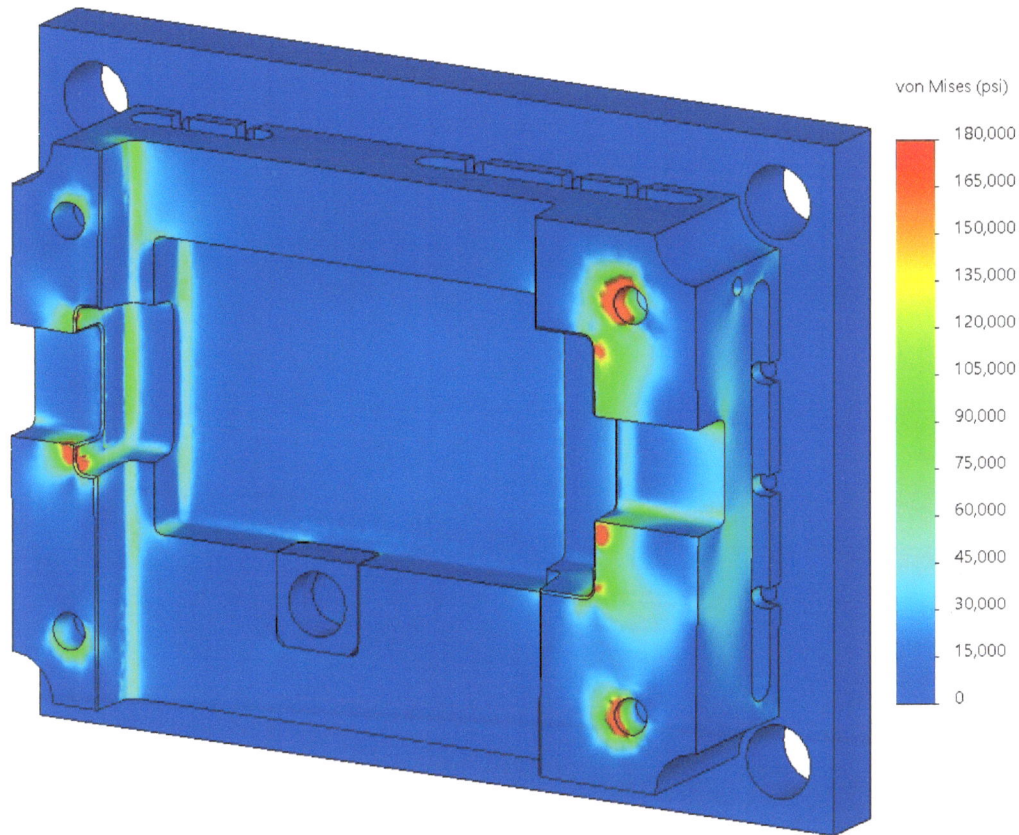

Result on cover holder

Overall results on the cover holder show that it is well stressed but not greatly more than in the service case (which was admittedly bad).

Here we have seen the power of contact analysis in a complicated interlocking assembly. We did not even scratch the surface of what detailed data can be pulled from such a model. Contact forces at every pair can be used to specify shimmed hardware and friction pads. Stress fields in each block can be used to guide placement of cutouts for cooling loops, ejection rods, and every other kinds of hardware. Finally, the whole die set can be shown robust (or not) to mis-adjustment or interference from trapped debris.

Simplified bell housing casting die
Two slides

Areas: *in^2*

257	projected (incl gating)
137	bell end
51.8	shaft end

Forces: *lbf*   14000 psi injection pressure

3598000	projected
1918000	bell end
725200	shaft end

1799 tons

			mesh dets	elements	run time	
1a	cover holder only	direct loading, fixed base	25.4/6.35/1.4	-	-	
1b	cover holder only	direct loading, bolted base (virtual wall)	25.4/6.35/1.4	100%	100%	
1c	cover holder only	contact loading, fixed base	38.1/9.5/1.4	108%	45%	
1d	cover holder only	contact loading, bolted base (virtual wall)	38.1/9.5/1.4	108%	227%	
2a	holders only	direct loading, fixed base	25.4/6.35/1.4	216%	223%	
2b	holders only	direct loading, bolted base (virtual wall)	25.4/6.35/1.4	216%	774%	
2c	add slides, inserts	pressure loading, fixed base	38.1/9.5/1.4	198%	179%	
2d	add platens, slides, inserts	pressure loading, fixed platen, tie bars	38.1/9.5/1.4	252%	1876%	equilibrium not achieved
3c	model change add rails	pressure loading, fixed base	38.1/9.5/1.4	214%	221%	
3d	model change add rails	pressure loading, fixed platen, tie bars	38.1/9.5/1.4	267%	1323%	
4c	model change add chip	pressure loading, fixed base	38.1/9.5/1.4	222%	287%	
4d	model change add chip	pressure loading, fixed platen, tie bars	38.1/9.5/1.4	275%	1605%	

Tabulation of runs

The summary of runs illustrates the growth in simulation solve time. Naturally, it took much longer to run all die pieces in full contact. Early simple studies showed that some parts of the response can be had from faster runs, but the great level of detailed data available from the full runs shows the power of a true and literal simulation setup.

Conclusion

Thank you for reading our book. We hope that in this guide the engineer or analyst has gained an expanded appreciation for what can be accomplished with fundamental simulation tools.

Contact Us

Contact Stone Lake Analytics with any questions or comments. We are a full-service design and analysis consultancy with specialties in casting dies, articulating machines, and welded structures. Models used in this book can be found at the company web site www.stonelakeanalytics.com.

About the editor
Shawn Mahaney is the lead analyst at Stone Lake Analytics. He has broad experience in design, testing, and analysis of products from automotive, freight rail, material handling, and casting process applications.

www.ingramcontent.com/pod-product-compliance
Lightning Source LLC
Chambersburg PA
CBHW041445210326
41599CB00004B/136